RADIO, TELEVISION AND AUDIO TEST INSTRUMENTS

Books on Radio, Audio and Television by Gordon J. King

F.M. RADIO SERVICING HANDBOOK

THE PRACTICAL AERIAL HANDBOOK

RADIO AND AUDIO SERVICING HANDBOOK

TELEVISION SERVICING HANDBOOK

COLOUR TELEVISION SERVICING

THE HI-FI AND TAPE RECORDER HANDBOOK

RAPID SERVICING OF TRANSISTOR EQUIPMENT

SERVICING WITH THE OSCILLOSCOPE

BEGINNER'S GUIDE TO RADIO

BEGINNER'S GUIDE TO TELEVISION

RADIO, TELEVISION AND AUDIO TEST INSTRUMENTS

GORDON J. KING
R.Tech.Eng., M.I.P.R.E., F.S.R.E., M.R.T.S., M.I.T.A.I.

LONDON
NEWNES—BUTTERWORTHS

THE BUTTERWORTH GROUP

ENGLAND	BUTTERWORTH & CO. (PUBLISHERS) LTD. LONDON: 88 Kingsway, WC2B 6AB
AUSTRALIA	BUTTERWORTH & CO. (AUSTRALIA) LTD. SYDNEY: 586 Pacific Highway Chatswood, NSW 2067 MELBOURNE: 343 Little Collins Street, 3000 BRISBANE: 240 Queen Street, 4000
CANADA	BUTTERWORTH & CO. (CANADA) LTD. TORONTO: 14 Curity Avenue, 374
NEW ZEALAND	BUTTERWORTH & CO. (NEW ZEALAND) LTD. WELLINGTON: 26–28 Waring Taylor Street, 1 AUCKLAND: 35 High Street, 1
SOUTH AFRICA	BUTTERWORTH & CO. (SOUTH AFRICA) (PTY) LTD. DURBAN: 152–154 Gale Street

Originally published, 1962, and reprinted 1963, as 'Radio and Television Test Instruments' by Odhams Press Ltd.

Second Edition published 1972 as 'Radio, Television and Audio Test Instruments', by Newnes-Butterworths, an imprint of the Butterworth Group

ISBN 0 408 00071 6

FILMSET BY FILMTYPE SERVICES LIMITED, SCARBOROUGH
PRINTED IN ENGLAND BY
HAZELL WATSON & VINEY LTD., AYLESBURY, BUCKS.

Preface

THIS HANDBOOK was written as a companion volume to supplement—but not overlap—my other books on radio and television servicing, some of which were originally published by Odhams Press Ltd., and are now republished by Newnes-Butterworths. This is the final book of those of the earlier series to go into a second edition and I must confess that I experienced greater difficulty with this volume than with any of the others. Much has happened in the world of instruments since the first edition was published and it was a tough job deciding just what to delete and what to add.

We have seen a dramatic change from valves to transistors and from analogue to digital readout. Moreover, we have witnessed the birth not only of an era of colour television, but also of a new audio age. For these reasons I have introduced two entirely new chapters, one dealing with colour television test instruments and the other with audio test instruments, both considering the application of the instruments involved.

I found it necessary severely to modify other chapters, some of which I have virtually rewritten with new illustrations and information on solid state techniques. Digital instruments are essentially (at present, anyway) outside our sphere of activity, but I have made reference to them where applicable. Indeed, some multirange testmeters are going digital and it is only a matter of time before they replace the analogue readouts in our workshops.

It has been my aim to keep the text at a level which would be appreciated most by the service technician, the student and the many thousands of experimenters who are for ever seeking to expand their knowledge. This is essentially a practical guide, and as such is written with the emphasis on application and testing methods. In some cases it has also been necessary to consider various design and circuit features; but this was done so that the reader would fully appreciate the scope and diversity of application of a particular instrument.

There must be a great many test instruments collecting dust simply because their users are not fully conversant with every possible use to which they may be put. This applies especially to the oscilloscope, and it is for that reason that a large part of this book is devoted to this remarkably versatile tool of the electronics engineer. I have endeavoured to indicate the limitations as well as the advantages of instruments and have also suggested ways in which certain tests may be undertaken with instruments seemingly designed for some entirely different purpose. Test instruments are costly, and it is economic folly not to exploit them fully. I sincerely hope that this book will at least result in more instruments being brought down off the shelf and put to good use!

Finally, my thanks are due to the manufacturers and my friends and colleagues in the electronics industry who kindly supplied circuit information, photographs and encouragement, without which this new edition could not have been launched.

Gordon J. King

Brixham, Devon

Contents

CONTENTS

CHAPTER 1

An Introduction to Instruments

THE SERVICE TECHNICIAN and the experimenter of today are in a much more favourable position with regard to test instruments than were their counterparts of the pre-television era. Even in the early days of television it was often necessary to compromise, the technician being obliged to carry out tests with a meter of dubious sensitivity and to place trust in the harmonics produced by an ordinary broadcast-frequency signal generator for diagnosis in television work. Today the situation is changed. The number of television receivers in use has increased from thousands to millions and from monochrome to colour; radio receivers have developed from those designed for ordinary broadcast frequencies with amplitude modulation, to those having v.h.f. circuits with frequency modulation; portable receivers have passed through the stage of valve operation and now employ transistors; while a new phase of sound reproduction has evolved, with hi-fi amplifiers, record players, control units and tape recorders.

Instruments have been produced capable of dealing with the new requirements of testing. Old designs have been brought up to date and instruments entirely new in principle have been developed. There is now available an instrument capable of making any test that may be necessary during the normal course of servicing radio, television and audio equipment.

The greatest requirements for modern servicing are accuracy and speed in the diagnosis of a fault condition. These can only be met by the technician or experimenter possessing a thorough knowledge of the fundamental methods of operation not only of the electronic equipment on which he is working, but also of the test equipment he will employ to help solve the problem for him. Moreover, the operator should be aware of every conceivable use that can be made of each item of his test equipment, for it sometimes happens that by adapting and combining methods of measurement and instrumentation a ready solution can be found to a servicing problem otherwise requiring a rather specialised test instrument.

TYPES OF INSTRUMENT FOR SERVICE WORK

Radio test instruments can be divided into three broad categories:

1. Those which indicate in absolute values and use an indicating pointer or digital readout. These include voltmeters, ammeters and ohmmeters along

with other devices which are designed and calibrated to give an indication of units such as those of inductance, capacitance, mutual conductance, magnification factor, and so on. They may either employ a meter movement, as does a voltmeter and a multirange testmeter, or a calibrated scale and control knob.

2. Those which produce a test signal either unmodulated or modulated with sound or television intelligence and which can usually be set to a specific frequency. These include signal generators, pattern generators, audio oscillators and generators and wobbulators.
3. Those which employ a cathode-ray tube as an indicator. These include the oscilloscope and other specialised items of test equipment.

Most service operations can be handled with an instrument from categories (1) and (2). For example, a sensitive multirange testmeter and a signal generator covering from, say, 110 kHz to 300 MHz would represent the basic requirements. There are a surprising number of service technicians who still fail to make full use of modern equipment and carry on with little more than a multimeter and a screwdriver. Such servicing techniques have much to commend them in the field, where it is virtually impossible to send out a quantity of complex test equipment, and the technician who is able to carry out successful repairs with little more than his sense of sight, hearing and touch is undoubtedly an engineer well on top of his job.

However, on the bench proper apparatus is essential if time is not to be wasted when the technician encounters tricky and unusual faults. It is desirable then to have substantially more than the two basic instruments mentioned earlier. The ideal service department includes the following equipment: standard high-resistance multimeter with automatic cut-out or overload protection; valve voltmeter; signal generator, amplitude-modulated; signal generator, frequency-modulated, with facilities for sweep-frequency operation; cathode-ray oscilloscope; pattern generator; audio oscillator or generator; electrostatic voltmeter or other means of measuring e.h.t. voltages; valve tester; transistor tester; capacitance bridge; picture tube tester and reactivator; miscellaneous items such as short-circuit turns tester, capacitor and resistor decade boxes, square-wave generator and so on.

Some of these instruments may be represented by the facilities provided by a single composite type of instrument. For example, the f.m. and a.m. signal generator may be integrated in a single unit, and similarly the pattern generator may be a facility given by the television signal generator. Integrated instruments of this nature have much to commend them for ordinary servicing work, for logical integration ensures that the instrument is always used to full capacity and is not left on the shelf to gather dust.

There is little doubt that a greater incentive is created for taking an interest in the finer points of servicing by the possession of or, at least, having access to reliable up-to-date instruments. New instruments mean that new servicing techniques have to be studied, and in general this also leads to a greater understanding of the equipment one is called upon to service.

Apart from instruments capable of generating signals at accurate very-high frequencies in the television and f.m. bands, the multirange meter is still a basic need of the service technician without which nothing much can take place, apart, perhaps, from a few sparks from the tip of a screwdriver! This test, of course, lets one know whether voltage is present or not, but it does not give an accurate idea of the amount of voltage.

The multimeter is really a personal instrument. One gets to know its characteristics and idiosyncrasies, if any—whether it reads slightly high on one range, low on a.c. and so on. There are a diversity of types, ranging from pocket meters with ten or so ranges, to large-scale instruments with fifty or more ranges.

Possibly the most important check in servicing is that for voltage, but curiously enough this is a difficult thing to measure accurately. Connecting the voltmeter makes it a part of the circuit in which the voltage is being measured. This alters the circuit constants and gives a wrong voltage reading. The amount of error, as will be discovered later, depends on the resistance or impedance of the test circuit in relation to the resistance or sensitivity of the voltmeter.

This trouble did not give rise to many problems in the early days of radio servicing when very high resistances were few and the voltage was not especially critical. Meters of relatively low resistance were then in use successfully. Today something far better is essential, since very high resistance circuits are used in television receivers and the connexion of a meter can detract severely from the normal operation of the set. Thus, it has been one aim of instrument manufacturers to produce multimeters with higher and higher sensitivites. Sensitivities have increased from something like 500 ohms-per-volt to 100,000 ohms-per-volt during the last decade. This is remarkable when it is considered that an instrument of 1,000 ohms-per-volt was looked upon as being highly delicate only fifteen years ago.

Although a sensitivity of 100,000 ohms-per-volt is bordering on that of the electronic voltmeter, this instrument is nevertheless required for many tests which demand the least amount of loading on the test circuits. An ordinary multimeter needs power from the test circuits to operate the meter movement, but with electronic voltmeters the power of the meter movement is provided by a valve or transistor, the supply circuits being within the instrument, and the only power required from the test circuits is the negligible amount needed for application to the grid circuit of the valve, or the gate or base of a transistor.

CHOICE OF INSTRUMENTS

As has already been intimated, the first great need to the experienced and inexperienced alike is a reasonably sensitive *multirange meter* of good quality. Such an instrument is essential for repairs on radio receivers, television sets, hi-fi amplifiers, tape recorders or, indeed, on any electronic device that the service technician will be called upon to repair. This instrument should thus be chosen with great care.

For ordinary radio work, a meter with a sensitivity of 1,000 ohms-per-volt is usually adequate, but for television work and more delicate operations on hi-fi equipment a great deal more can be done with a meter whose resistance on d.c. voltage ranges is at least 20,000 ohms-per-volt (full-scale deflection requiring 50 microamperes). Generally speaking, the higher the sensitivity the less robust are the meter movements, but this should not be taken to mean that meters with very high sensitivities are hyper-delicate and not mechanically suitable for the service department. This may have been true several years ago, but today the scale of sensitivity is increased by the use of high-efficiency magnets and finely wound moving-coils, and even a meter with a sensitivity of 100,000 ohms-per-volt will survive for many years in the service department with no more than the usual care which should be given to all test instruments.

Some multirange meters read only d.c. volts and current, while others read d.c. volts and current and a.c. volts. However, the larger type of instrument also has facilities for reading alternating currents, and these are more expensive. For general use, a meter providing for measurements of a.c. and d.c. volts and current is highly desirable, for then heater-current readings are possible and also audio output-power tests.

Such instruments usually provide about eight d.c. volt ranges, seven or eight a.c. volt ranges, seven to ten d.c. current ranges, and four to eight a.c. current ranges. In addition, resistance measurement ranges are provided. The highest resistance reading depends on the sensitivity of the meter movement and on the voltage of the internal battery, but instruments of 20,000 ohms-per-volt sensitivity usually measure up to 20 megohms, while an instrument of 100,000 ohms-per-volt would measure up to 200 megohms with its internal battery. The resistance ranges can be extended by the use of an external battery of greater voltage than that contained within the instrument. The lowest resistance measurement depends on the design of the instrument; the average multirange meter indicates 10 ohms or so, but some indicate as low as 0·5 ohm.

The range of voltages should be studied with care to ensure that they suit the applications to which the instrument will be subjected. It should also be noted that, although the a.c. volts sensitivity relies upon the sensitivity of the meter movement, it is affected to a large degree by the bridge rectifier and associated components used in the instrument. This usually means that the sensitivity on a.c. volts is rarely as high as that on d.c. volts. Typical examples are 20,000 ohms-per-volt d.c. and 1,000 ohms-per-volt a.c., or 100,000 ohms-per-volt d.c. and 5,000 ohms-per-volt a.c. These ratios, of course, differ between instruments of various manufacture.

High current can be measured equally as well on low and high sensitivity meters, but it is possible to measure much smaller currents on high sensitivity meters than on low sensitivity types. The highest a.c. current range provided really depends on the cost of the instrument, for to measure current requires current shunts and rectifiers which can be switched from one range to another, and this is usually an expensive exercise.

Multimeters also sometimes measure inductance, capacitance, decibels and so on either directly or by the use of adaptors. Adaptors are also available to permit the measurement of very high voltages, such as television e.h.t. voltages, but these are only suitable for this application if the instrument has a d.c.

sensitivity exceeding 20,000 ohms-per-volt.

Some form of automatic cut-out or overload protection is a desirable feature, for no matter how careful a technician may be, under the pressure of great thought there is always a possibility of his connecting a meter wrongly; for example, a meter adjusted to read 'ohms' may inadvertently be connected across the h.t. line. Unless there is some kind of protective device, this inadvertent action will almost certainly seriously damage the instrument.

For workshop use, the electronic voltmeter (using valves or transistors) has much in its favour. The meter movement need not be as delicate as those employed in highly sensitive conventional multirange meters, yet the overall sensitivity is far in excess of even that of the most sensitive multirange meter. These instruments have a very high input resistance, and thus reflect very little load on the test circuit. They are admirable for making checks in the a.g.c. circuits of radio and television receivers, where the circuit impedance is very high, for acting as an alignment indicator in f.m. discriminator and ratio detector stages, and for tests in the timebase, video and colour circuits of television sets.

On a.c. voltage their loading to the external circuit is little more than on d.c., and their frequency response is far in excess of that of multimeters switched to a.c. volts. By the use of an external probe, signal frequencies as high as 100 MHz can be measured with reasonable accuracy. With an ordinary multirange meter a response up to about 10 kHz is all that can be expected. Valve voltmeters can thus be used for measuring the stage gain of r.f. and a.f. amplifiers. Such an instrument is essential for serious work on hi-fi audio equipment.

Multirange versions of the electronic voltmeter or 'electronic testmeter', as it is sometimes called, are available and have ranges almost comparable to the conventional multirange meter. However, because of their very high input resistance, very high d.c. voltage ranges are often incorporated. Some models can be switched to 1,000 V, 3,000 V or even 5,000 V, and by the use of an external multiplier the range is extendable to 10,000 V or 20,000 V. Apart from its other attributes, therefore, the electronic voltmeter is very useful in television servicing because it provides a means of measuring the e.h.t. voltage at the final anode of the picture tube.

Multirange electronic voltmeters also in general have a wide range of resistance scales. The highest resistance which can be measured may be in excess of 1,000 megohms, which is sufficiently high to be useful for insulation tests, while the lowest resistance range may cater for the accurate measurement of a fraction of an ohm.

Like their passive counterparts, some models may also measure capacitance, inductance, power output and decibels. A.C. ranges give useful indication of 100 mV, but millivoltmeters (those measuring only volts and indicating decibels) often give full-scale deflection on an input as low as 100 μV.

Choice of *signal generator*, which is the next most important instrument, will depend on the type of work to be undertaken. As a general-purpose instrument, one covering from, say, 100 kHz to 240 MHz, with facilities for amplitude-modulation, is desirable, as it can be used for the alignment of ordinary broadcast receivers as well as for some television sets, and can also be used to provide signals in Bands I, II and III, for checking front-ends and overall sensitivity. For Bands IV and V a u.h.f. output is required.

Signal generators with facilities for both amplitude modulation and frequency modulation have a distinct advantage but all models of this kind do not cover the broadcast bands, which means that a separate instrument is required for the alignment of ordinary radio sets. There is at least one service-type instrument which features two independent oscillators and tuning scales, for a.m. and f.m. The instrument also contains facilities for sweep alignment, which, in conjunction with an oscilloscope, enables the response curve of the tuning circuits to be displayed on the cathode-ray tube. Thus, any alteration to the tuning of the set under test is revealed in terms of response variation. When the instrument is so employed, the a.m. oscillator can be used to provide a marker pip on the displayed response curve.

Some f.m. or a.m./f.m. signal generators provide for sweep alignment, and when employed in this way are often known as 'wobbulators'. These instruments either provide a signal for operating the X timebase of the associated oscilloscope, or the sweep is effected by the X timebase signal of the oscilloscope. The sweep-frequency range is also important, especially if it is required to align visually the vision i.f. stages of television receivers. For television, a maximum sweep of at least 10 MHz is desirable, but for a.m. and f.m. radio receivers, a maximum sweep of 1 MHz is sufficient.

Signal generators in a slightly higher price group may feature a crystal oscillator for checking the instrument's calibration frequency, a meter for monitoring and setting the r.f. signal as applied to the attenuators, and a modulation depth meter.

Transistor signal generators are now current and useful in locations where mains power supply is difficult. They are portable, and being powered from internal batteries can immediately produce a test signal in a customer's home. The valve-type instrument, of course, must be connected to a mains supply and this is not always so easy, as every field service technician knows.

Like the valve instruments, transistor signal generators are made in quite a wide range of prices, from fairly inexpensive units for amateur and servicing use to expensive and complex models designed specifically for alignment and setting-up of battery-operated communication equipment and laboratory use.

Transistors already have almost ousted valves in radio test instruments. This is not because valve type generators are in any way inferior, but portability can be an asset in the laboratory or service workshop.

Other signal generators include those designed specifically to produce a television carrier on which can be modulated a series of video waveforms for creating patterns on the screen of a television receiver under test. This type of instrument is known as a 'pattern generator'. Pattern generators are often capable of providing useful information concerning the overall performance of television sets, and certain models almost simulate the information contained in a test card transmission. They can, therefore, be used to set up television receivers from scratch without an aerial signal.

There are two classes of pattern generators: a rather simple kind in which the

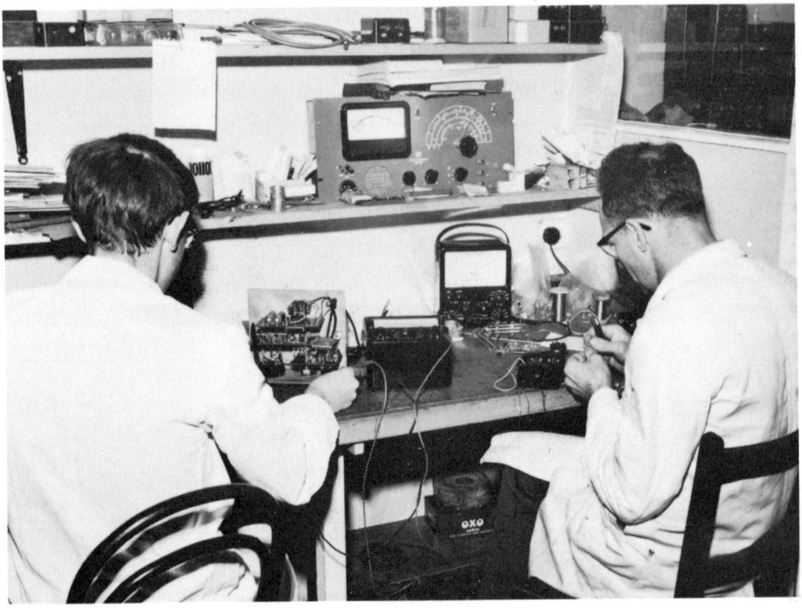

FIG. 1.1. *Corner of a manufacturer's servicing department, showing engineers at work repairing v.h.f. amplifiers and test equipment. Instruments visible include multirange testmeter and transistor tester (Taylor Instruments) and v.h.f. signal strength meter (J. S. Fielden). (Transistor Devices Ltd., South Devon.)*

carrier is modulated by square-wave signals, which produce vertical and horizontal dark bars on the screen, and a more complex arrangement which, in addition to various patterns, produces a line and frame synchronising waveform for locking the receiver's timebases, as does an aerial signal. In effect, the latter kind produces a video waveform closely simulating that sent out by the transmitter. This type of instrument is by far the best, of course, but is considerably more costly than the simple arrangement.

With the advent of colour television a new type of pattern generator has emerged. In its simplest form this produces dot and crosshatch patterns to facilitate adjustments to the static and dynamic convergence controls. There is also a more complicated instrument of this kind which yields colour-encoded signals in addition to the dot and crosshatch patterns to the NTSC or PAL (or both) standard, giving a display in the standard colour bars or similar hues.

Although most signal generators provide, in addition to the r.f., an audio signal at a fixed frequency, usually 400 Hz, an audio oscillator or generator is of marked value to the technician likely to be concerned with the repair of audio equipment. The instrument should supply a good-quality sine wave over the frequency range of, at least, 20 Hz to 30,000 Hz. It should feature a low output impedance, in the region of 600 ohms, and a reasonably accurate attenuator which will control the signal level from below 1 mW to 1 watt, at least.

Some instruments of this kind embody a meter movement calibrated in volts a.c. as a means of checking the level of the output signal. Others also provide

facilities for square-wave production, which can be very useful for distortion and transient tests on hi-fi amplifiers. In general, an audio generator allows for signal injection tests, frequency response measurements, and distortion tests of power amplifiers, control units, tape recorders, public address installations, loudspeakers and like equipment.

One way and another, there are so many signal generators of various types and at various prices available that the amateur constructor or service technician choosing one for the first time may find it difficult to decide which to select. Of course, the most expensive instruments will give all possible facilities and are desirable if price is no object. But buying an expensive instrument for routine work may be rather like buying an expensive precision camera in order to take family snapshots. If one is not going to use all the refinements on the generator, one might as well save money and settle for a less expensive unit with just the facilities needed.

A useful generator at modest cost might cover a.m. only, from the lower medium frequencies (say 100 kHz) to the top end of Band III (240 MHz). One would have to pay more for the addition of f.m. and sweep alignment facilities, but to many service engineers these will be essential.

It is also worth paying extra to have an instrument where the required bands are covered exclusively on fundamental frequencies instead of having to rely on harmonics to obtain top-band coverage. The tuning-scale arrangements are important and should make for easy reading, particularly on the high-frequency bands, since interpolation is often necessary to establish a tuned frequency between two calibrated points on the scale. A vernier dial helps in this respect, but is usually found only on more expensive generators.

It should also be remembered that the *minimum* signal that a signal generator can apply to equipment on test depends upon how much signal 'escapes' from the instrument by radiation. In the more costly models special attention is paid to the screening of the oscillator and attenuators to avoid excessive radiation, and an overall screen in addition to that provided by the metal case is always fitted. These precautions keep stray radiations to very small levels.

The ordinary experimenter or service technician may feel that such highly efficient screening is not really necessary so far as his work is concerned. However, if a good deal of stray signal is picked up by sensitive equipment under test from the generating case, then effectively the attenuator is by-passed when switched to high attenuation ratios. This makes it almost impossible to perform tests involving sensitivity measurements on highly sensitive equipment.

For example, if the sensitivity of a receiver is 5 microvolts and the signal picked up by the set from direct radiation is 20 microvolts, then it is virtually impossible to obtain an absolute sensitivity value. A stray field due to radiation of about one microvolt on the v.h.f. bands would represent an excellent specification for a general-service type of instrument. Laboratory instruments may have a specification of less than 0·1 microvolt, and this is one reason why such instruments may be so much more expensive.

One of the most versatile of test instruments is the *oscilloscope*, and once its basic operation is understood it will be employed nearly as much as the multi-

meter. Apart from the production of waveforms, it can be used to measure a.c. and d.c. voltage with negligible loading on the test circuit; it can act as a signal tracer, frequency comparator, waveform analyser, and is very useful for fault tracing in all types of electronic equipment, especially in the timebase circuits of television receivers.

There are many test applications for which an oscilloscope is particularly suitable; some of these only the oscilloscope can handle and the others it can often handle better than any alternative instrument. The oscilloscope should be employed for routine tests, as well as for diagnosing complex faults. If it is brought into use only when all else has failed, then its full value will never be appreciated.

The minimum performance specifications required depend on the kind of equipment likely to be tested or repaired. However, as a general guide the cathode-ray tube should be between 3 and 5 inches in diameter, the vertical or Y amplifier should have a sensitivity of at least 100 mV/cm, with its frequency response flat within 3 dB from d.c. (or at least 20 Hz) to 2 MHz. For television work, a larger bandwidth and sensitivity are desirable. Some relatively inexpensive instruments have a sensitivity of 30 mV/cm from 10 Hz to 8 MHz. The timebase should cover from 10 Hz to, at least, 100 kHz with trace expansion, or to 500 kHz if trace expansion is not provided.

Other things to look for are: viewing hood and graticule ruled in centimetre squares for calibration and assessment of applied signal amplitude; polarising filter; tilting stand; X terminals for operating a wobbulator; triggered and repetitive timebase synchronising.

Other Test Equipment

In addition to the basic instruments described earlier there are available many other possibly more specialised, instruments designed particularly for the service technician and experimenter. These include valve, picture tube and transistor testers, signal tracers, resistor and capacitor boxes, output meters, electrostatic voltmeters for testing television e.h.t. voltages, and many more which cannot be described in detail in this chapter, but which are investigated in subsequent chapters.

It is worth noting that certain instrument manufacturers are tending more and more to integrate various instruments in a single housing. Where one's financial situation has a bearing on the purchase of instruments, the integrated types are well worth investigating, as it may be found that two or more instruments can be purchased in this way for not much more than the price of one.

Kit instruments are also becoming popular, an exercise having educational as well as economic value.

CONSTRUCTION OF INSTRUMENTS

The meter movement most often found in radio testing equipment is the moving-coil, permanent-magnet type. Such a movement consists of a circular or

horseshoe-shaped permanent magnet with its pole pieces forming opposite sides of a cylindrical opening. In this opening is supported the moving-coil unit which is wound in the form of a square with a very fine gauge of wire on a former made of aluminium or copper foil. The pointer, made of non-magnetic material, is attached to the coil, to which is secured two pivots. In order to provide the requisite sensitivity, the whole assembly must be extremely light.

The moving-coil with its pivots is mounted between two bearings, either of white metal, or of sapphire in first-class instruments. The coil is thus free to rotate within the field of the permanent magnet and, in order to enhance the field strength surrounding the coil windings, a cylindrical iron core fills most of the space between the pole pieces. Lines of force pass through the coil and core from north pole to south pole of the permanent magnet, and when current flows through the coil the field of the coil reacts against the field from the permanent magnet and subjects the coil to a rotating force, as shown in Fig. 1.2.

This rotating force is opposed by the force exerted by two hairsprings, one at each end of the coil. The inner end of each control spring is connected to one end of the moving coil, and the outer ends are attached to connecting tags which, in turn, connect to the movement terminals. The springs are so mounted that

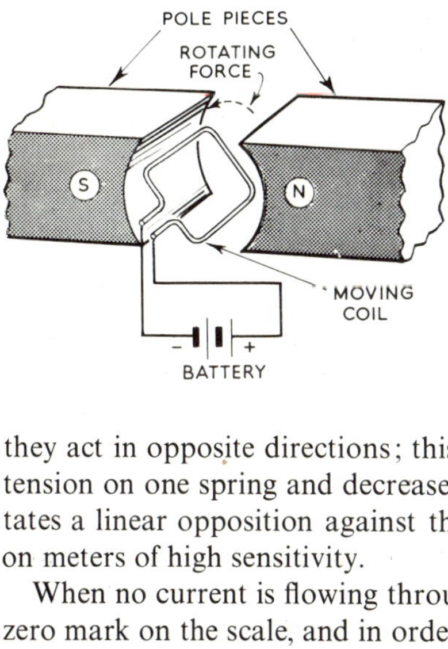

FIG. 1.2. *The basic principle of the moving-coil meter. The pole pieces form the ends of a permanent magnet. A cylindrical core, not shown, fills most of the space between the pole pieces, leaving an annular gap in which the coil rotates.*

they act in opposite directions; this is to say, rotation of the coil increases the tension on one spring and decreases that of the other. This arrangement facilitates a linear opposition against the electromagnetic effect, which is essential on meters of high sensitivity.

When no current is flowing through the meter, the pointer should rest at the zero mark on the scale, and in order for this to be accomplished the tags which are connected to the outer ends of the springs can be rotated to increase or decrease the tension. The front tag can usually be moved slightly on either side of vertical by means of the 'zero adjuster' on the front of the instrument case. A small pip set eccentrically at the rear of the adjuster engages in a slot in the tag, and rotation causes a side-to-side movement of the tag, shifting the pointer a certain amount either side of the zero mark on the scale. By opening the case, the rear tag can also be moved, allowing the pointer to be set to zero when the front adjusting screw is in its mid-position.

Some instruments include automatic temperature compensation in the form

of a bi-metal structure to overcome the effects of temperature variation. The entire moving assembly must be perfectly balanced, and this is achieved by small balancing weights secured to the spindle. A magnetic shunt, made of a piece of soft iron, pivoted at one end on one of the pole pieces of the magnet, can be adjusted to open and close the magnetic field and thus modify the sensitivity of the movement. This should not usually require adjustment, but is set at the factory to endow the movement with the required full-scale deflection and overall sensitivity. There are two delicate springs to check the movement of the pointer at the extremes of the scale.

Moving-coil instruments are available for the measurement of current over a wide range of values. Full-scale sensitivities of 20 to 50 microamperes are commonplace these days, while sensitivities down to 10 microamperes are available for more specialised applications; 1 mA movements are by far the most common and feature extensively in instruments designed for general service work. Moving-coil instruments indicate correctly over the scale only when the current is flowing in one specific direction through the coil. If the current is reversed, the pointer will kick against the zero stop. However, instruments are available with a centre zero, and these are useful for the balancing of bridge circuits and discriminators. One of the greatest advantages of a moving-coil instrument is that the deflection is linear with respect to the flow of current in the coil. Thus the scale calibration is uniformly spaced.

Moving-Iron Meter

This type of movement is generally less sensitive than its moving-coil counterpart, and instead of a moving-coil uses a small soft-iron armature. This is mounted at the centre of a fixed coil, through which the current to be measured flows. Deflection is produced by the interaction of the field of the coil and the magnetism induced in the soft-iron armature. Unlike the moving-coil meter, however, the moving-iron meter will function irrespective of the polarity of its terminals, and thus is suitable for the measurement of alternating currents and voltages. Note that the pointer deflection does not bear a linear relationship to the applied current, and this results in a progressive widening of the scale divisions from zero to full-scale.

By the use of various shunts and series resistors, meter movements can be arranged to measure over a wide range of voltage and current, and by the inclusion of a battery in a voltmeter circuit the movement can be scaled to measure resistance. Such additions in conjunction with a moving-coil meter form the basis of the multirange test set, which is dealt with in greater detail in subsequent chapters.

Valve-Voltmeter

The heart of this instrument is the moving-coil meter connected across a Wheatstone bridge arrangement. One arm of the bridge is formed by a d.c.

amplifying valve, and the other arms are made up of resistive networks. A 'set zero' control enables the bridge to be balanced so that the meter indicates zero when the instrument is disconnected from a signal or voltage source. When a voltage is applied to the grid of the valve, the balance of the bridge is disturbed and the meter reads in proportion to the value of the applied voltage.

For the measurement of audio and high-frequency signal voltages a diode rectifier, which is often contained in the a.c. probe of the instrument, is introduced between the signal input and the grid of the valve. If a diode probe is used, the loading on the signal circuits can be very much reduced, since the major part of the probe lead then simply carries the rectified signals to the valve grid. The range switching may be carried out in the resistive arms of the bridge, and the scale of this type of meter is usually calibrated in r.m.s. input voltages according to the ranges covered. The inside of the Avo Electronic Testmeter is shown in Fig. 1.3, where the a.c. probe is clearly visible on the right-hand side. Below frequencies of approximately 10 Hz, the signal loading effect is not so great and the probe can be operated inside the instrument, in which case the terminals make contact with two metal strips and the signal voltage can be applied in the usual way to the front panel terminals of the instrument.

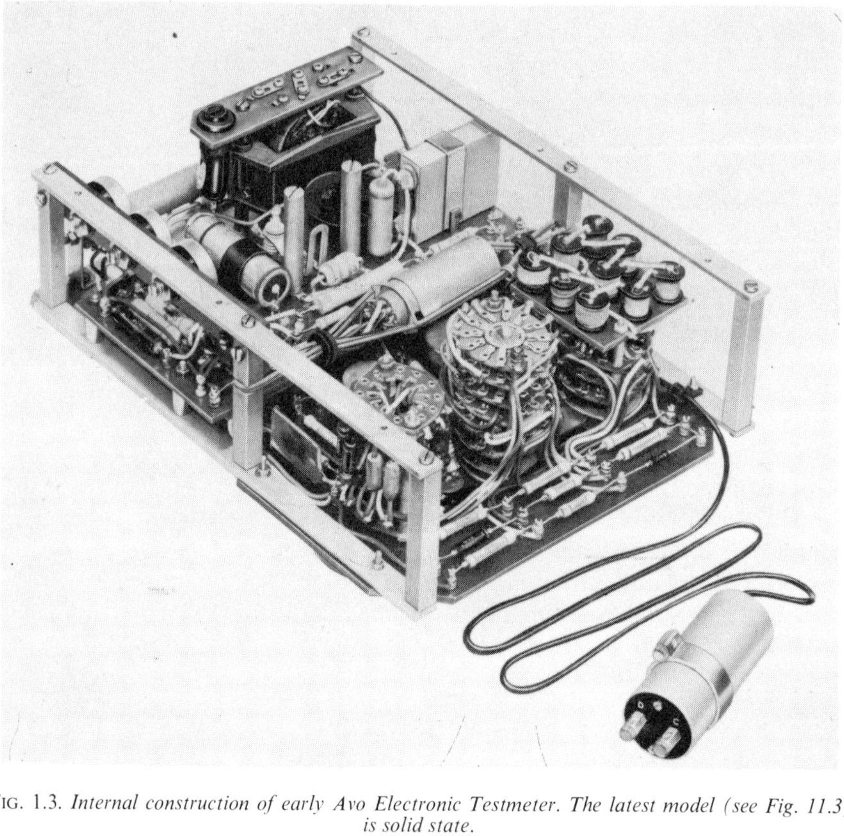

FIG. 1.3. *Internal construction of early Avo Electronic Testmeter. The latest model (see Fig. 11.3) is solid state.*

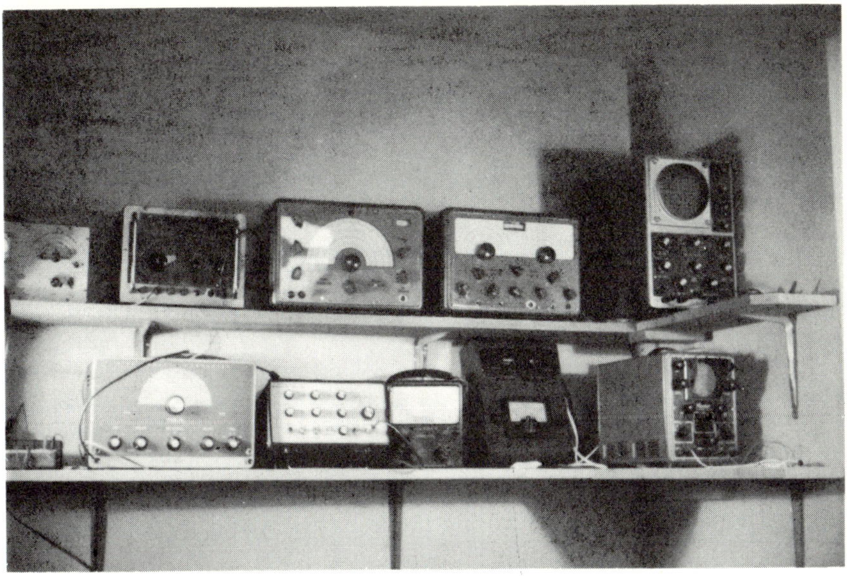

FIG. 1.4. *Corner of the author's studio showing several radio, audio and television test instruments.*

Transistors are rapidly replacing valves in this type of instrument, and in this guise the instrument is known either as an 'electronic voltmeter' or transistor voltmeter. It is not uncommon to find multirange instruments in this category, using transistors and working from internal batteries. The advantage of portability can be appreciated, especially by those engineers and technicians who in the past had to locate a mains supply to operate the valve voltmeter.

Audio Millivoltmeter

Again this is an instrument which employs either valves or transistors but it differs from the valve or transistor voltmeter or multirange meter in that it is designed to measure a.c. signals in the audio spectrum and often above from a few Hz up to at least 30 kHz over a voltage range down to a few millivolts (sometimes microvolts) and up to units, tens, hundreds or thousands of volts. An instrument of this kind is essential in audio servicing and testing, it being required in particular for checking the frequency response, power and distortion of audio amplifiers, and for these purposes a decibel scale (or scales) is commonly included.

Signal Generators

Under this heading is included the Modulated Test Oscillator, which is in effect a second-class signal generator primarily intended for the test bench. The standard signal generator is virtually the same kind of instrument with a much better performance and standard of accuracy. However, most signal generators which are available to the service technician today have specifications which a few years ago would only apply to much more expensive equipment.

The signal generator consists of an r.f. oscillator which is tunable and capable of producing a carrier within a specific range of frequencies, an a.f. oscillator which serves to modulate the carrier, an attenuator which allows adjustment to the modulated r.f. output voltage and a dummy aerial which feeds the generator

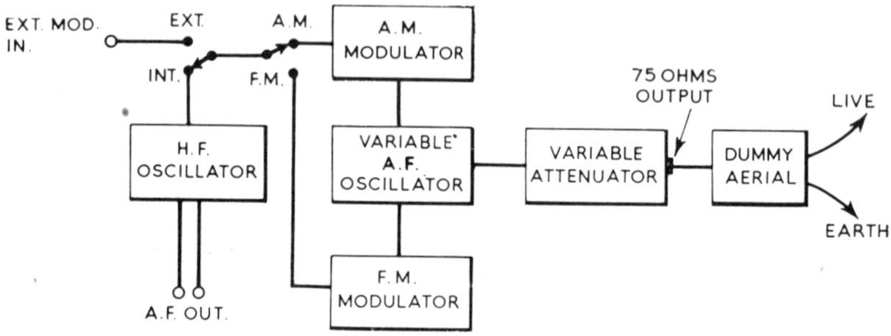

FIG. 1.5. *Block diagram of signal generator for a.m. and f.m.*

signal to the receiver under conditions similar to those of an ordinary radio aerial. Most modern instruments also provide an r.f. output at 75 ohms to match the input of television receivers. Facilities for both frequency modulation and amplitude modulation are often provided, and the general arrangement of an f.m./a.m. instrument is shown in the schematic diagram of Fig. 1.5.

The r.f. oscillator may be a conventional Colpitts circuit and the a.f. oscillator a resistive or inductive feedback arrangement. It is also generally possible to modulate the r.f. carrier from an external a.f. source. A step attenuator, in 20 db intervals, and a continuously variable attenuator control allow a smooth range

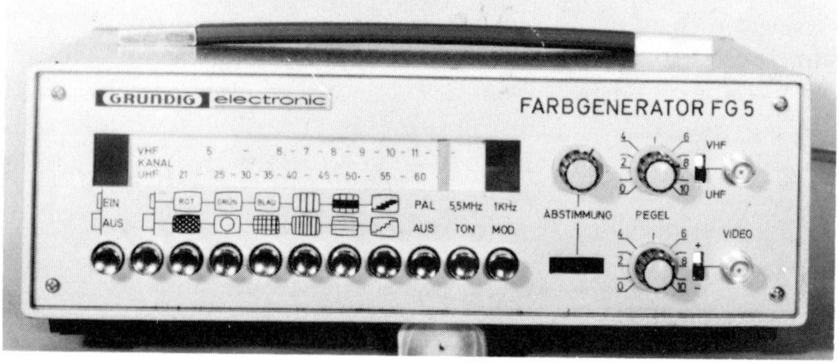

FIG. 1.6. *Colour television service generator based on the PAL standard. (Grundig Electronics)*

of r.f. output voltage adjustment. These controls may be calibrated in terms of voltage output, the usual range being from 1 microvolt to 100 mV, or decibel values.

Audio Generators

The usual circuitry in this type of instrument is a resistive-capacitive feedback circuit using two triodes and an output pentode. The required frequency coverage is provided by switched and variable resistors, and it is possible to obtain a good-quality sine wave with a power output of approximately 1 watt, matched to two or three load resistances.

More recent instruments are based on transistors in a Wien-bridge circuit and low-distortion amplifier. Low distortion is essential in order to check the distortion performance of a hi-fi audio amplifier without the signal source itself adversely affecting the readout and without having to filter the signal in an endeavour to remove unwanted harmonics.

Most instruments designed for audio applications also incorporate a circuit for changing the pure sine wave to a square wave for the investigation of transient response and amplifier stability into reactive loads. For this purpose the rise time must be fast, preferably less than 0·5 μs.

The circuit of an excellent yet inexpensive instrument of this kind (Model Si 453 by J. E. Sugden) is illustrated in Fig. 1.7. This has a frequency coverage of 13 Hz to 30 kHz in six overlapping ranges (avoiding calibration cramping), a maximum output of 2 V r.m.s., output stability of ±0·2 dB, total harmonic distortion less than 0·05% at 1 kHz and a square wave rise time less than 0·5 μs. A very useful feature is the switchable RIAA filter at the output. This delivers a signal conforming to the fine groove recording characteristic of B.S. 1928 as would be obtained from a magnetic gramophone pickup playing a test record recorded to this characteristic. Output is 10 mV maximum at 1 kHz and adjustable down to zero. This facility allows the RIAA equalising of a hi-fi amplifier to be checked in terms of a 'flat' output.

Another popular sine wave/square wave generator is shown in Fig. 1.8. This is the Heathkit IG-82U, the firm now named Heath (Gloucester) Ltd., formerly Daystrom. It is available either in kit form or for a little extra fully constructed and tested.

A somewhat different arrangement of obtaining audio signals is by means of the beat-frequency oscillator. This type of instrument uses two oscillators, one set at a fixed frequency and the other variable, both operating considerably above audio frequency, around 100 kHz. The two signals are mixed and the frequencies are arranged so that the resultant beat frequency, which is the difference between the two oscillator signals, gives the required audio signal. The signal from the mixer passes through a low-pass filter and thence, via a gain control, to an a.f. amplifier, as shown in the schematic diagram of Fig. 1.9.

This type of oscillator produces output signals from about 20 Hz to 10 or 12 kHz, whereas the audio generator usually covers from 20 Hz or below to about 20 kHz, or even much higher than this.

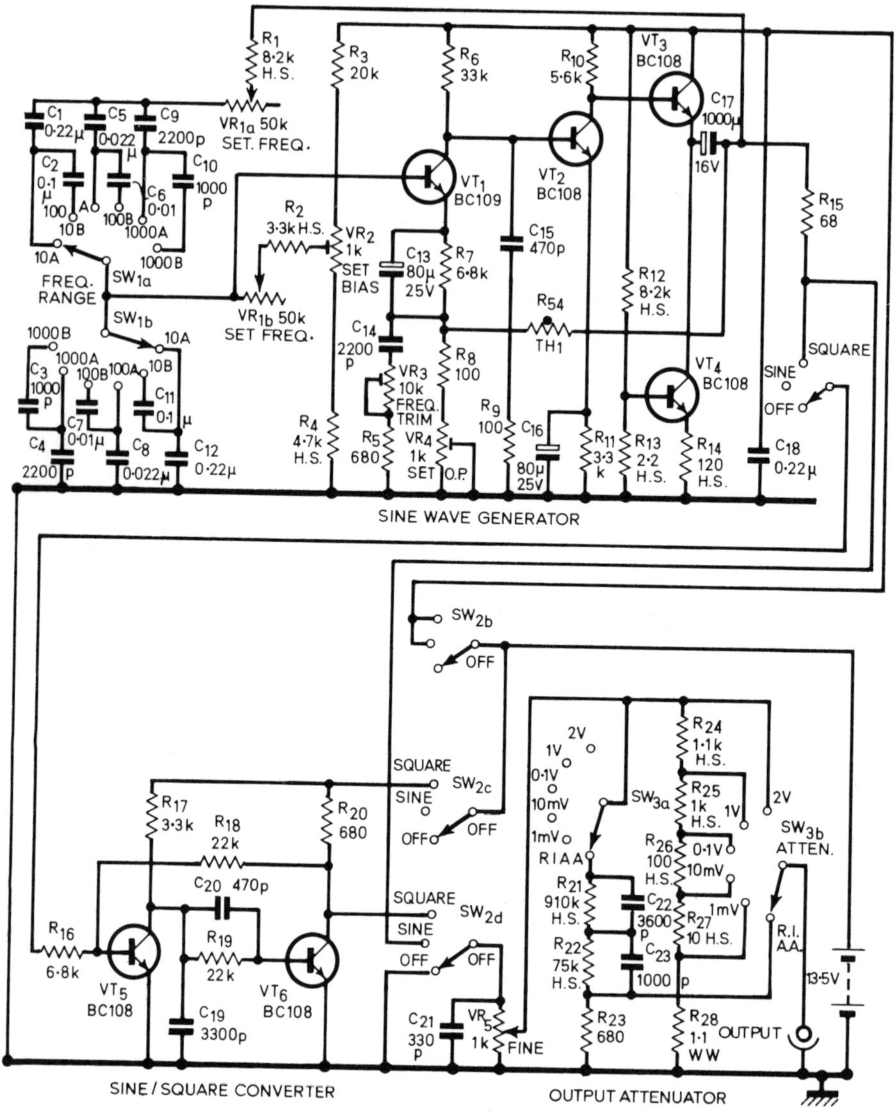

FIG. 1.7. *Circuit diagram of audio oscillator by J. E. Sugden, Model Si 453. The sine-wave generator is based on a Wien-bridge circuit and the design is such that the waveform distortion is typically less than 0·05% at 1 kHz. The thermistor in the feedback circuit yields an output stability of ±0·2 dB. The generator can either be switched direct to the attenuator or through a two-transistor sine/square converter to produce a square-wave output of rise time less than 0·5 μs. Fine and coarse attenuators are included and there is also a switched RIAA filter so that the output simulates that from a magnetic pickup cartridge playing a frequency test record recorded to the RIAA standard.*

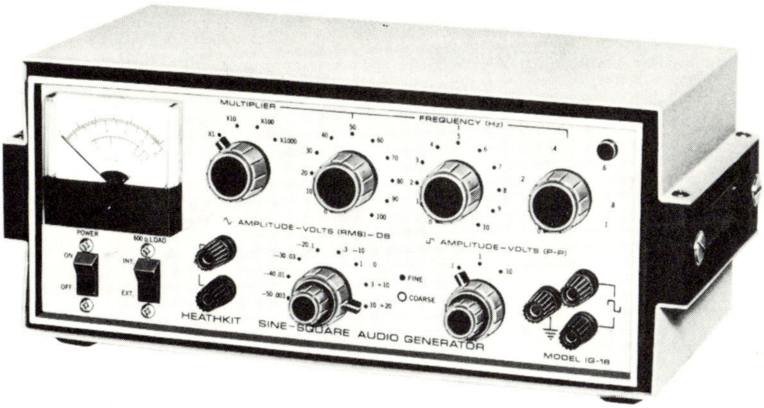

FIG. 1.8. *Sine wave/square wave audio generator by Heath (Gloucester) Ltd, Model IG-18. This features separate controls for the sinewave and squarewave outputs, with separate output terminals.*

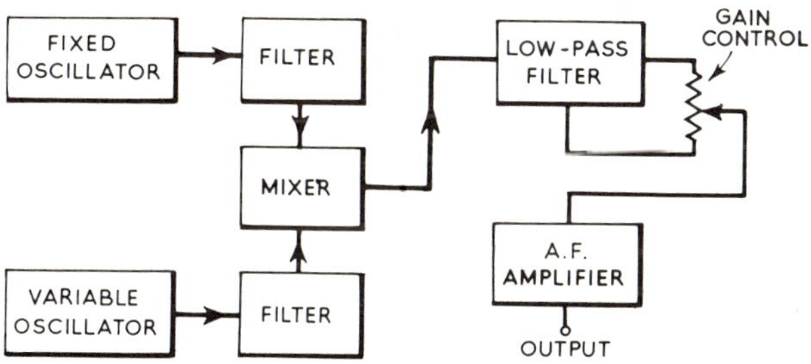

FIG. 1.9. *Block diagram of beat-frequency oscillator. The signals from the two oscillators are mixed to produce a resultant beat frequency, which provides the audio signal.*

Oscilloscopes

Basically, the oscilloscope consists of a cathode-ray tube, vertical deflection amplifier, linear timebase for horizontal deflection and a power supply. Some instruments also contain a horizontal, or X, amplifier for providing trace expansion. The output of the timebase or X amplifier is fed to the horizontal deflecting plates of the tube, which causes the spot produced on the screen of the tube to move steadily from left to right across it. At the end of its travel, the spot is caused to 'fly back' to its starting point. This is accomplished by the sawtooth waveform which the timebase produces.

The number of scans per second can be controlled, and the range may be from one per second, or even below this, to 5,000 or more per second. At low scanning speeds the movement of the spot can be followed by the eye, but above about 15 the spot can no longer be seen and a horizontal line appears on the screen, whose flicker effect disappears above about 30 scans per second.

The signal to be examined is applied to the Y plates of the tube, via the vertical or Y amplifier if greater deflection sensitivity is required. In this way the waveform of the applied signal voltage is traced on the screen by the spot. For example, if a 50 Hz sine wave, from the mains supply is applied to the Y plates and the timebase is adjusted for 50 scans per second, then one complete waveform of the signal would be displayed on the screen.

In order to hold the timebase in step with the Y signal, a part of the Y signal is fed back to the timebase to synchronise it. This ensures that the display remains steady, since it may otherwise tend to drift to the right or to the left of the screen even though the timebase frequency has been carefully adjusted to suit the frequency of the test signal.

Some oscilloscopes have twin or split beams, in which two horizontal scanning lines are produced and which can be individually controlled. In such instruments two Y amplifiers are used as well as separate horizontal and vertical shift controls for each beam.

For colour television work the Y amplifier should have a −3 dB bandwidth over at least 4·5 MHz (to embrace fully the colour subchannel components in and around the chroma amplifier and decoder). Very convenient oscilloscopes are now being made for battery or mains operation, based on transistorised circuits. These are useful for field servicing in particular owing to their portability. More and more colour television servicing work calls for in-the-field diagnosis, amounting to the tracing and identification of complex waveforms at the various test points and printed circuit boards and modules. Indeed, today an oscilloscope is just as important as the multirange testmeter was in the heyday of radio (see, for example, the author's books on *Colour Television Servicing* and *Servicing with the Oscilloscope*, by the same publisher).

Distortion Test Sets

An essential instrument for audio servicing is the distortion analyser. In its simplest form this merely 'notches out' the fundamental, leaving a power summation of all the generated harmonics. These are commonly measured in terms of r.m.s. *voltage* across the output load with an audio millivoltmeter and referred to as percentage distortion relative to the steady-state r.m.s. voltage across the same load resulting from the power in it. Thus distortion voltages 40 dB below the voltage resulting from the fundamental power in the load would represent a total harmonic distortion (THD) of 1 % since 40 dB is a 100 : 1 voltage ratio. The actual *power* of the distortion components in the same dB ratio would, of course, be 10,000 times below the *power* of the fundamental since a 40 dB power ratio is 10,000 : 1. Hence an amplifier tested at, say, 10 W with a THD of 1 % would yield 100 mV of r.m.s. distortion voltage across a 10-ohm load and 1 mW of distortion power in it. This is because the power in watts is equal to the *square* of the voltage divided by the load resistance in ohms or to the *square* of the current multiplied by the load resistance in ohms.

THD is commonly expressed as:

$$\% = \frac{\sqrt{A_2^2 + A_3^2 + A_4^2 \ldots}}{\sqrt{A_1^2 + A_2^2 + A_3^2 + A_4^2 \ldots}} \times 100$$

where A_1, A_2, etc, are the voltage or current values of the fundamental and harmonics.

A more complicated distortion measuring instrument is the wave analyser. This is in effect a very selective voltmeter with a tunable narrow passband filter. The filter is tuned to the harmonic component of the fundamental to be measured and a meter, based on an audio millivoltmeter circuit, reads the selected harmonic distortion in terms of the r.m.s. voltage of the harmonic across the load relative to the r.m.s. voltage of the fundamental signal.

Distortion testing instruments can also be used to measure intermodulation distortion with two a.f. signals. The signals chosen are commonly 100 Hz (f_1) and about 5 kHz (f_2) and f_1 is usually arranged to have a definite ratio to the amplitude of f_2, a ratio of 4:1 being conventional. Non-linearity in an amplifier results in signal f_2 being modulated by signal f_1 and its harmonics, thereby giving rise to intermodulation products at frequencies of $f_2 - f_1$, $f_2 + f_1$, $f_2 - 2f_1$, $f_2 + 2f_1$, $f_2 - 3f_1$, $f_2 + 3f_1$ and so on.

Intermodulation distortion (IMD) is commonly expressed as:

$$\frac{\sqrt{(A_{f2-f1} + A_{f2+f1})^2 + (A_{f2-2f1} + A_{f2+2f1})^2 + (A_{f2-3f1} + A_{f2+3f1})^2 + \ldots}}{A_{f2}} \times 100$$

Here $Af_2 - f_1$ refers to the r.m.s. voltage (or current) of the component resulting from the intermodulation at the frequency $f_2 - f_1$ at the output load of the amplifier.

Measurements of the IM products may be made individually and the distortion calculated from the above formula or they may be made by filtering the IM products and measuring their r.m.s. value.

At the measured IMD level the corresponding output *power* can be calculated from the following expression:

$$\text{Output power} = \frac{(V_1 + V_2)^2}{R_L} \text{ or } (I_1 + I_2)^2 \times R_L \text{ watts,}$$

where R_L is the load resistance in ohms and the subscripts refer to the frequencies f_1 and f_2.

Digital Instruments

Some digital voltmeters for measuring d.c. are based on a voltage-to-frequency converter followed by a digital counter. The converter produces a train of

FIG. 1.10. *Digital multimeter, Type DMM2 by Advance Electronics Ltd. This employs both transistors and integrated circuits. The block diagram in shown below.*

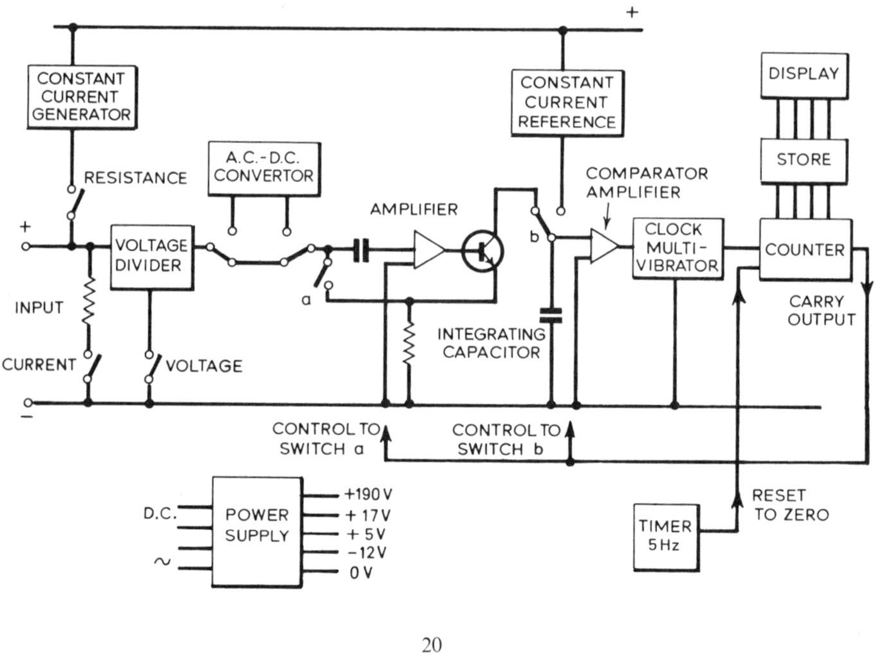

pulses which is linearly proportional to the input voltage and the counter totals the number of pulses yielded during a specific period of time. Thus it becomes possible to design a circuit, taking into account the parameters of pulse frequency and count period, whereby the digital readout is equal to the value of the d.c. input voltage.

Very clever circuit configurations are adopted in instruments of this kind and remarkably high readout accuracy is achieved. It is also possible for a digital instrument to act as a 'counter' for frequency measurements. Digital instruments are obviously many times more complicated than simple analogue meters and are thus much more costly. All recent models are based on transistors and/or integrated circuits and semiconductor devices of various species. Some models incorporate more than 100 transistors to provide the required circuit functions, and the design is not uncommonly based on plug-in printed circuit board 'modules', each module handling a certain function. For example, there may be modules for the input circuit and amplifier, voltage-to-frequency converter and pulse generator, timer counter, display counter, first decade, second decade, third and final decades, control circuits and power supply. In spite of the complexity and large number of circuit elements, both passive and active, the use of semiconductor devices ensures that the instruments are of quite reasonable size.

At the time of writing digital instruments are employed mostly for computer maintenance and for design activities, but since their cost is on the decrease they will eventually find their way into less exotic laboratories and servicing workshops and ultimately into the homes of radio, television, audio and electronic enthusiasts.

CHECKING INSTRUMENTS

Most radio and television test instruments can be checked fairly easily for fundamental operation by using them for trial tests and measurements on equipment which is operating correctly, but of the kind that the instruments will later be used to service. For example, a signal generator can be set up to feed a modulated signal into a television receiver. Depending on the channel setting of the receiver, the signal generator when tuned to the vision channel frequency—and in this way the instrument's calibration can be roughly judged—will cause a series of dark, horizontal lines to appear on the screen. The lines are due to the modulation signal, and the number of lines will depend on the modulation frequency and the setting of the receiver's frame hold control.

If the receiver is in good order and reasonably sensitive, this pattern should be well in evidence with the instrument's attenuator adjusted to give an output of, say, 100 microvolts. When the modulation is switched off the screen should display a raster which is in general brighter than when the r.f. signal is switched off, disconnected or detuned.

Tuning to the sound frequency of the channel selected with the modulation on, a tone should be heard from the loudspeaker. The same method of checking can be adopted with ordinary a.m. or f.m. sound receivers. In addition, a fairly accurate calibration check is possible by injecting the signal when the set is also receiving an aerial signal. The generator's signal should be fed to the aerial socket via a 'hold-off' resistor and the modulation switched off. The 'hold-off' resistor can be about 75 ohms.

When the generator is tuned so that its output frequency approaches the frequency of the received transmission, a whistle, varying in pitch as the tuning is altered, will be heard from the loudspeaker. The whistle will start at a very high pitch and, with very careful tuning of the generator, will decrease in frequency to zero (known as the 'dead-beat' point) and then increase in frequency again at the other side of zero or dead beat. When the generator is tuned to dead beat, then the frequency of the signal it is producing is exactly the same as that of the transmission, and since this will be known, the amount of calibration error of the generator can be read off the scale. The idea is shown in Fig. 1.11, where at (*a*) is shown the receiver generator set-up and at (*b*) the dead beat point.

When this check is used on ordinary radio sets, the signal should be injected via the dummy aerial; the 'hold-off' resistor is only necessary for f.m. and television receivers having a low-impedance aerial input circuit. With portable receivers, a small length of wire should be connected to the 'live' terminal of

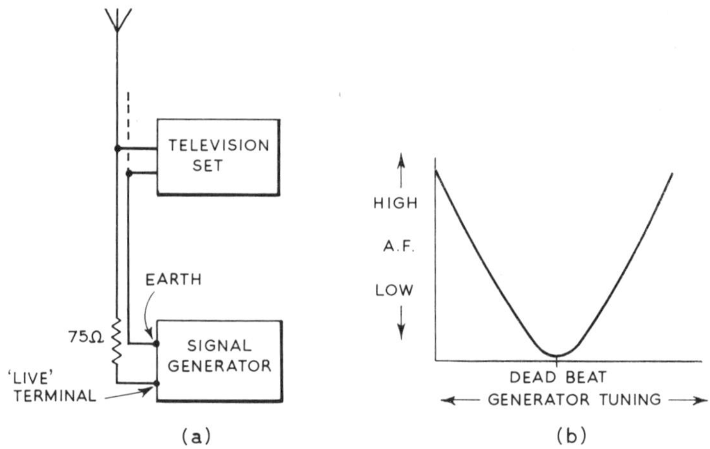

FIG. 1.11. *Calibration check of a signal generator.* (a) *Television receiver and signal generator set-up for making a frequency check.* (b) *Graph showing the 'dead beat' point which occurs when the generator frequency coincides with the frequency of the off-air signal.*

the dummy aerial and this should be positioned close to the receiver's frame or ferrite rod aerial. It will be necessary to adjust the strength of the injected signal to secure the best overall results. The calibration of the generator can be checked over its various bands against off-air signals of known frequency.

Current meters and voltmeters can be checked against known current and voltage values in a receiver circuit, for instance. An ordinary dry cell can also give a useful indication of the accuracy of the low-range reading of a voltmeter.

But it must be remembered that a single dry cell, when new, will read slightly in excess of the expected 1·5 volts when not supplying current.

The higher a.c. voltage ranges of a multimeter can be checked against the mains voltage, but an accurate readout is not possible by this method owing to the tendency for local mains voltages to deviate from the nominal voltage in these days of heavy electricity demands. The lower ranges can be checked against the low-tension voltage across the heater windings of receivers' mains transformers, but, again, an accurate check is almost impossible. The current ranges can be checked by metering the current from a battery of known voltage through a resistor of known value. The current reading should equal the voltage divided by the resistance. The alternating current in a series-connected heater chain provides a reasonably accurate source with which to check the low and intermediate ranges.

The 'ohms' ranges can be checked reasonably well with a number of radio resistors of assorted values. The 'set zero' control should first be carefully adjusted, and if a reading well past full-scale deflection cannot be achieved by rotating the 'set zero' control fully clockwise with the test leads shorted, the internal battery should be replaced.

With larger instruments and, indeed, with even the simplest of instruments, the initial checks should be made in conjunction with the instruction manual. This usually contains the basic information necessary to perform simple tests, assuming that the operator is completely without knowledge of that type of instrument.

With an oscilloscope, for example, the instruction manual will contain information as to the correct setting-up procedure, the operation of the various controls and their effect on the displayed waveform, methods of measurement, and so on. It is not intended in this chapter to duplicate such information, but subsequent chapters will take over where the instruction manual leaves off (also see the author's *Servicing with the Oscilloscope*).

It is often possible to check instruments in relation to each other. For example, both an oscilloscope and a signal generator could be checked by feeding the generator's signal into the Y amplifier of the oscilloscope. Assuming that the Y amplifier of the oscilloscope has a bandwidth which embraces the applied r.f., then such things as modulation depth, r.f. output voltage, envelope distortion and so on can easily be checked, thus proving that both the signal generator and the oscilloscope are correct.

CALIBRATION

The calibration accuracy of instruments very much depends on the purpose for which they are designed. Voltmeters, ammeters, multimeters and signal generators designed specifically for the service technician usually have an accuracy of the order of 2 per cent. Some signal generators in the popular price group have accuracies of the order of 1 per cent. With a crystal check facility, a signal generator will have an accuracy at the check points of something like

0·005 per cent, which is the fundamental accuracy of the crystal itself within a specified temperature range. It is thus possible to make appropriate corrections to the scale based on the check points.

Accuracy is defined as the maximum error that is likely to occur, as measured in a 'plus or minus' percentage. With indicating meters this is a percentage of the full-scale reading. Therefore, the meter reading may be the specified percentage higher or lower than the voltage or current applied to the meter, but only at full-scale. For example, a meter with a full-scale deflection of 100 V and an accuracy of 2 per cent could give a reading between 98 V and 102 V when a potential of exactly 100 V is applied across its terminals. The *full-scale* accuracy still applies at smaller readings, so that if, say, 50 V is applied to the meter the reading may fall between 48 and 52 V, which at this point on the scale represents a possible plus or minus 4 per cent error. At even smaller readings the error will be greater; a quarter-scale deflection will represent a possible error of plus or minus 8 per cent, and so on.

It is for this reason that the ranges of test meters should be chosen with care so that the near full-scale readings will apply to the majority of measurements that the instrument will be called upon to make. If great accuracy is demanded at a potential of, say, 10 V, and the meter reads 100 V full-scale, it would be pointless to employ a meter with an accuracy of less than 0·1 per cent, for a meter with 1 per cent accuracy is likely to give a 10 per cent error at 10 V. It would be far better to use a meter with a 10 or 15 V full-scale deflection of lower accuracy.

Laboratory instruments are made with accuracies as good as 0·5 per cent and 0·1 per cent, but these are delicate and very costly devices. For ordinary service work an accuracy of 2 per cent is adequate. Such a meter allows general measurements to be made to show whether a particular section of the circuit is operating correctly, as distinct from very accurate measurements of voltage and current. The chief considerations from the service technician's point of view are robustness and versatility, and neither of these factors is possible with very delicate instruments. A meter with an accuracy of, say, 2 per cent is likely to remain this way in spite of occasional knocks, but this cannot be said for laboratory instruments of high accuracy, which may well fail totally under the normal service department treatment.

The accuracy of a.c. meters which employ an internal rectifier is usually below the d.c. accuracy. This is because of various errors which occur in the rectifier, and these decrease the accuracy to about 5 per cent of full-scale deflection with a meter which would have an accuracy of 2 per cent on the d.c. ranges.

PARALLAX ERROR

This refers to the error which is likely to occur when taking a reading when the line of sight is not perpendicular to the pointer and scale. The diagram in Fig. 1.12 shows how the apparent reading deviates from the true reading when the scale is viewed at an angle.

In order to eliminate this error, some instruments are fitted with an anti-parallax mirror on the scale plate. The correct reading is that indicated when the image of the pointer falls directly behind the pointer itself. This, of course,

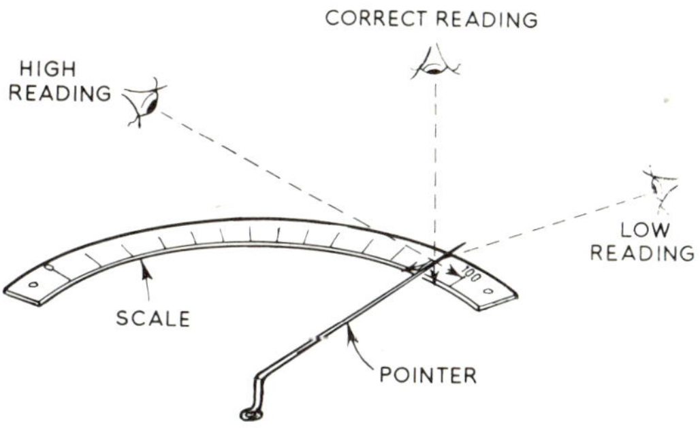

FIG. 1.12. *Illustrating parallax. To obtain a true reading the line of sight must be perpendicular to the pointer and the scale.*

simply ensures that the reading is taken when the line of sight is exactly perpendicular to the scale.

Similar errors can occur on all pointer-and-scale type instruments, including the signal generator and the oscilloscope, particularly where the latter is used in conjunction with a graticule to measure amplitude or time, or where a marking on the graticule is used as a datum line to which the trace is set.

CARE OF INSTRUMENTS

Although instruments designed for general service work are extremely robust, they must, nevertheless, be handled with reasonable care if they are to maintain their accuracy and usefulness over a long period of time. Instruments used for work in the field, such as multirange meters, signal strength meters, and pattern generators, should preferably be transported in a padded box which itself should be conveniently secured inside the service van. The box should be divided into sections to fit snugly the various instruments carried.

Some multimeters have specially padded cases, and these are very useful and can be recommended for outside service work. Excessive vibration can quickly destroy the accuracy of instruments, especially those which contain tuned circuits, such as signal generators and signal strength meters. Moving-coil instruments are less prone to this trouble, but a bad jolt may damage the pivots or the bearings.

Moving-coil instruments, particularly those housed in non-metallic cases, should always be kept clear of strong magnetic fields such as may emanate from loudspeakers, chokes and transformers. Although an external field of this nature may be insufficiently strong to damage the meter, it may affect the reading. If there is any doubt about the accuracy of a reading when used close to equipment containing such components as mentioned above, it is prudent to make further measurements with the meter in different positions, as a check against magnetic field influence.

Should it become necessary to remove a magnetic-type instrument from its case, care should be taken to ensure that the bench and surrounding tools are free from iron filings or small pieces of metal. Once an iron particle finds its way into the magnet system of such an instrument it is extremely difficult to extract it. Remember that the tools which are used to dismantle the instrument may be slightly magnetised and may be carrying minute particles of metallic dust, and this is likely to be attracted by the greater field of the instrument's magnet.

Never be tempted to alter any trimmer on an instrument in an endeavour to secure improved accuracy at one particular frequency, as this could put the instrument severely out of adjustment at other frequencies. Such alignment should never be attempted without reference to the maker's instruction book or service manual.

Instruments should always be stored in a dry and well ventilated place to avoid internal corrosion. Mains-operated instruments should be connected to the power supply and switched on for about a quarter of an hour at least once a week. To avoid damage in the workshop, meter flyleads should not be left trailing across the bench or on the floor. Multirange meters should always be left switched to the highest voltage position.

SAFETY PRECAUTIONS

Many radio and television receivers of recent manufacture adopt the so-called a.c./d.c. technique. This means that the heaters of all the valves are connected in series, and then applied across the power supply via a voltage-dropping resistor. With this arrangement, half-wave rectification is used for obtaining a high-tension supply, again direct from the power supply, and the chassis of the receiver is connected to one side of the power supply, which corresponds to h.t. negative.

It is essential, therefore, before making any tests or measurements, to ensure that the chassis is connected to the neutral side of the power supply. By far the most desirable precaution, however, is to connect the receiver to the power supply via a 1-to-1 ratio isolating transformer. This will completely isolate the power supply from the receiver whichever way round the power is connected. The receiver chassis may then be bonded to an efficient earth point. Under this condition, test instruments can be connected to the receiver without the risk of damage due to an unexpected power supply return circuit. Moreover, such a transformer will substantially minimise the risk of electric shock to the operator.

If an isolating transformer is not used, then certain instruments will have to be connected to 'a.c./d.c.' equipment via isolating capacitors. This applies in particular to signal generators and instruments which are used to inject a signal into the receiver circuits. Details of such isolation are given in a later chapter.

When making voltage tests on an unfamiliar circuit, it is always prudent to set the meter to a position which is far in excess of the expected voltage, and then tap the meter down, step by step, until the meter indicates at least half-scale deflection. An accurate reading may then be taken.

In high-voltage circuits, the negative lead should be firmly secured, usually to the chassis, and the positive probe can be connected to the required points with the right hand while keeping the left hand in one's pocket (or vice versa). This procedure is essential when making tests of e.h.t. voltages in television receivers.

When making current tests, always first switch off the equipment under test and insert the meter leads in the circuit in which it is required to measure the current. Make sure that the positive lead is connected to the positive side of supply (see Fig. 1.13), set the meter to a high current range and then switch on.

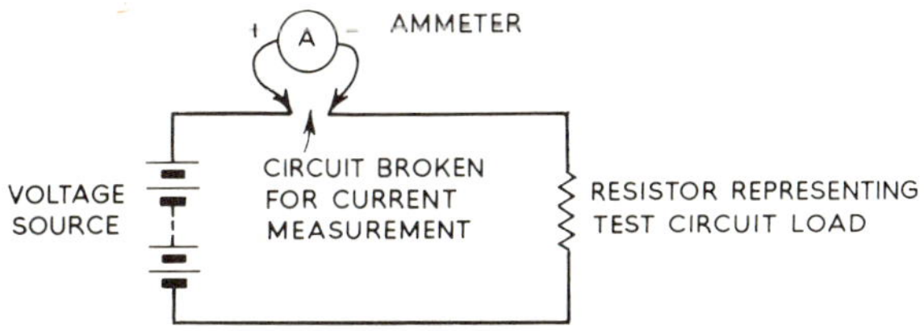

FIG. 1.13. *When current is to be measured, see that the ammeter is connected with correct polarity. The circuit should be broken and the ammeter must then be inserted with its positive terminal connected to the positive terminal of the supply source, as shown above. If the meter is to be connected at the other side of the load, the ammeter's negative terminal should then be connected to the negative terminal of the supply.*

Drop down on the current range until at least half-scale deflection is obtained, depending on the current ranges of the meter, but *always* disconnect the meter from the circuit, or switch off the circuit, before changing to another range.

If a testmeter is severely overloaded there is always a possibility that damage may result to the moving-coil mechanism and pointer, to the moving-coil itself or to the internal resistors in the case of a multimeter. Such damage is more likely when the instrument is overloaded on the current ranges, for then very heavy current may flow through the moving-coil, causing it to overheat and burn the insulation on the wire, or even causing the wire to fuse. In the voltage positions the current through the moving-coil is somewhat limited by the series resistors.

Overload Protection

Under the pressure of great thought even the best of us is capable of connecting a meter set to a low voltage range across the h.t. line of a receiver. Instrument manufacturers are fully aware of this and have devised methods of protecting instruments in this event.

Avo, for example, have employed mechanical automatic overload protection for many years. On the front of the instrument this appears as a small 'cut-out' knob. In the event of an overload, this knob springs from its normal position and breaks the meter circuit. The circuit is restored and the instrument made ready for further use simply by depressing the knob. The mechanism is brought into play by the moving-coil coming into contact with a trigger just beyond its full-scale position. There is, in addition, a second release at the zero end of the scale, so that the meter is likewise cut out if overloaded in reverse (see Fig. 1.14).

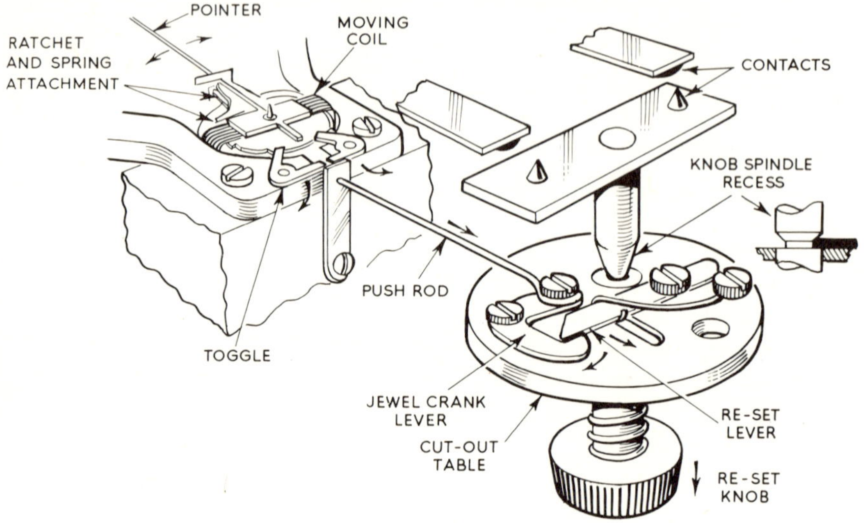

FIG. 1.14. *The Avo automatic cut-out mechanism. In the event of an overload, the pointer operates a toggle as it passes beyond full-scale deflection. The toggle moves a push rod and this in turn a crank lever, which releases the re-set lever. The knob spindle then springs free and opens the circuit contacts.*

Protection of meter movements is not uncommonly provided by diodes which effectively 'switch on' and thus by-pass overload current when the normal range voltage has been exceeded.

Transistorised instruments and those using valves have a 'built-in' safety margin resulting from the bottoming of the active device when the range input is exceeded. Of course, it is possible to destroy the active device, especially when this is a transistor, by the application of an input in advance of the junction capabilities and by the inadvertent introduction of high amplitude transients.

ACCESSORIES FOR INSTRUMENTS

Once the basic instrument is obtained, there are various ways by which its usefulness can be increased. Many instrument manufacturers produce accessories for use with their instruments, and while these may not be required immediately, it is as well to keep in mind such items as are available for any specific instrument. For example, with most high-resistance multirange meters adaptors are available for extending the voltage range, allowing the instrument to be used for measuring e.h.t. voltage in television receivers. In some cases, such adaptors take the form of a well-insulated probe in which are a series of high-voltage resistors of a total value sufficient to extend the range to 30 kV.

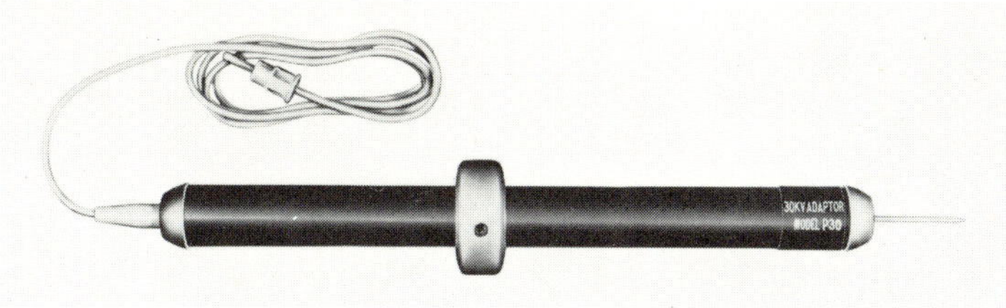

FIG. 1.15. *For the measurement of e.h.t. voltages in television sets a meter reading up to at least 30 kV is required. Here is shown a high-voltage probe for stepping up the range of a multirange test-meter to secure a full-scale deflection of 30 kV. (Salford Electrical Instruments Ltd.)*

It will be appreciated, of course, that such adaptors could not be used with medium- or low-resistance instruments, since the power required to operate the meter movement would be far in excess of the power capabilities of an ordinary television set, from the e.h.t. point of view. Meters with a sensitivity of 20,000 ohms/volt or more are suited to these adaptors.

Adaptors containing current transformers of a size greater than is possible to build into an ordinary multirange meter, for extending the current range, are also available from several manufacturers. Similar accessories are available for valve voltmeters, including r.f. probes which may or may not be sold as a part of the instrument. As has already been intimated, r.f. probes reduce losses and loading on the test circuit by allowing the test signal to be applied direct to a diode which is incorporated in the tip of the probe. Thus, the test signal is rectified and converted to d.c. before it can be attenuated or distorted in connecting cables or test leads. Moreover, the effect of the test leads on the test circuit is isolated by the diode. Adaptors are also available to connect to multirange meters to permit the measurement of capacitance and inductance. Signal generators have such accessories as dummy aerials, 75-ohm isolated and non-isolated terminations and crystals at various frequencies to provide calibration check points.

Accessories for the oscilloscope include an electronic switch for trace doubling. This extends the application of a single-beam oscilloscope by allowing it to provide simultaneous traces of two separate signals. In other words, it provides almost the same facilities as a double-beam oscilloscope. Two traces are often very useful, for they allow the direct comparison of two signals. For example, when checking for distortion in an audio amplifier, the radio test signal actually applied to the input of the amplifier can be monitored by one trace, while the output signal can be observed by the other. Distortion occurring in the amplifier is then immediately apparent.

More for laboratory work, although often very useful to the experimenter, is the oscilloscope camera. This provides a means of recording oscilloscope traces, either by taking single shots of a stationary waveform or continuous shots of non-recurrent waveforms and slow transients, on a 35 mm film. Some of the oscillograms (as film records of oscilloscope traces are sometimes called) reproduced in later chapters of this book were so taken by the author.

CHAPTER 2

Application of D.C. Meters

WITH THE EXCEPTION of digital and electrostatic meters, the indication given by the pointer in all direct measuring instruments is due to the flow of current through the coil or coils in the meter movement. Even when voltage is being measured, the meter is really indicating in terms of current, although the calibration is in voltage. Current is, therefore, required to be drawn from the test circuit in order to work the meter, and problems can occur when this current is insufficient.

A meter movement is designed to have a particular sensitivity; that is, the amount of current required to deflect the pointer from zero to full-scale. All moving-coil meters respond only to direct current. The application of a.c. to such a meter would simply produce a rapid change in magnetic field around the winding and cause the pointer to vibrate in proximity to the zero mark. In order to measure a.c. with such an instrument, a rectifier is required between the applied voltage and the movement to convert the a.c. to d.c. This problem is considered in the next chapter.

CURRENT MEASUREMENT

An ordinary meter movement with a sensitivity of, say, 1 mA will give full-scale deflection when 1 mA of current is flowing through the moving-coil. Since the scale is linear, it will give half-scale deflection when 0·5 mA is flowing, and deflections in proportion to the current down to about 0·1 mA or 100 microamperes. If smaller current is required to be measured, then the only solution lies in the use of a meter of greater sensitivity.

However, it is well worth bearing in mind that certain meters calibrated at about 10 mA full-scale deflection may, in fact, be of greater inherent sensitivity and employ an internal shunt which is designed to by-pass a proportion of the current applied across the meter terminals. Such meters sometimes originate from ex-Government sources, or may be taken from a piece of redundant electronic equipment. Take care when removing a shunt of this kind, since it may not be included solely to decrease the meter sensitivity. In some cases a form of shunt resistance is included, usually in the form of a coil of copper wire, to serve as a swamp resistance to the moving-coil proper to improve the temperature coefficient of the instrument. After a shunt has been removed the meter will, of course, need to be recalibrated and its full-scale sensitivity checked against a current of known value.

To measure currents greater than the full-scale sensitivity of the basic movement, it is necessary to use a shunt. This is, in effect, a resistance of very low value which is connected in parallel with the moving-coil, across the meter terminals, as shown in Fig. 2.1. As has already been mentioned, its purpose is to by-pass a proportion of the total current, so that only a fraction flows through the meter winding.

A shunt does not impair the accuracy of a movement, because a fixed relationship always exists between the resistance of the meter and that of the shunt. The shunt resistance (R_s) can easily be worked out for any meter from the expression

$$R_s = R_m/n-1$$

where R_m is the internal resistance of the meter and n is the factor of multiplication. Take for example a 1-mA meter with which it is required to read 100 mA full-scale. In this case the factor of multiplication is 100, and if the meter has an internal resistance of, say, 100 ohms, which is a fairly common value, then we shall have 100/100 − 1, or 100/99, which works out to 1·01 ohms.

Shunt resistors invariably have a very low value which may, without a resistance bridge, present somewhat of a problem to measure. The shunt should

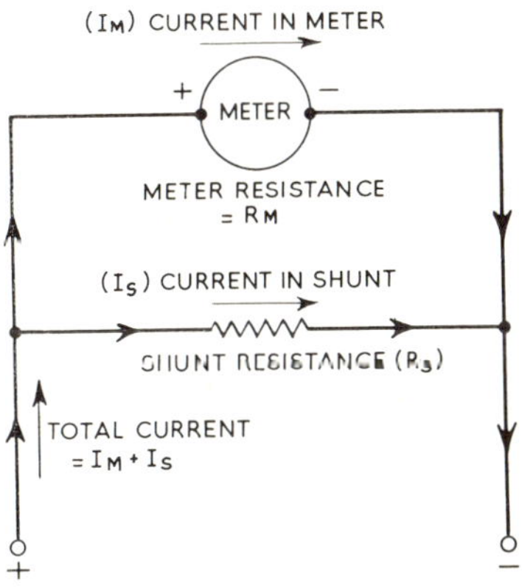

FIG. 2.1. *Currents greater than those for which a meter is designed can be measured by the inclusion of a shunt resistance* (R_S) *connected across the meter terminals, as shown. The total current is then divided so that the greater part will flow through the shunt and only a computed proportion flows through the meter.*

be capable of withstanding the by-pass current without overheating and should be fixed securely inside the instrument (or outside if the shunt is large or is required to pass a large current—in the region of amperes). It should be remembered, of course, that an increase in temperature of the shunt will result in an increase in resistance, with a consequent error in the meter reading. Copper wire strapped between the two terminals, or formed as a self-supporting coil, would be suitable for a shunt of very low resistance. Manganin is also satisfactory.

When the approximate resistance has been ascertained from wire tables, in accordance with the required shunt value as calculated, the shunt can be checked by setting up a circuit as shown in Fig. 2.2. Slight alteration in length of the shunt wire can be made until both meters indicate the same current. Practical shunts can be made either of Eureka or nickel-chrome resistance wire wound on a Paxolin former. To avoid having to disconnect the shunt wire when making an adjustment, a little extra wire can be wound on the former

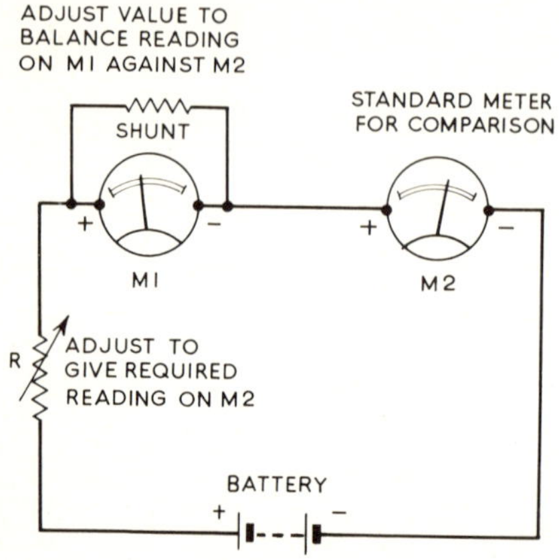

FIG. 2.2. *Using a standard meter as a comparison to adjust the shunt resistance on a second meter. The resistor R should have a maximum value to give approximately half-scale deflection on M2 in relation to battery voltage. For example: total R = E/I, where R is in ohms, the battery potential E is in volts, and I is the half-scale reading of M2 in amperes (0·001 ampere = 1 mA).*

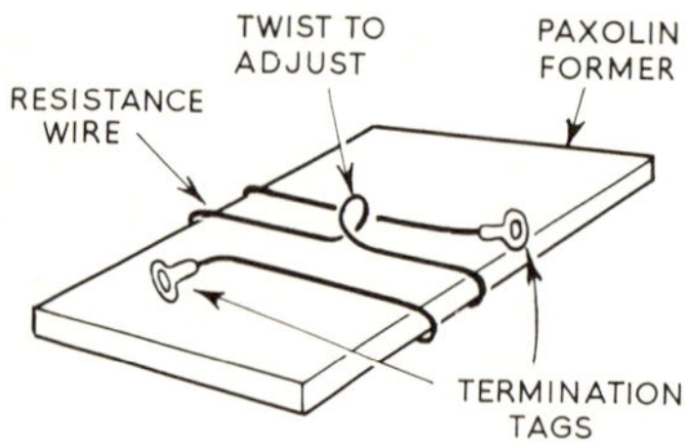

FIG. 2.3. *A practical shunt resistance. Fine adjustments can be made by twisting or untwisting the loop.*

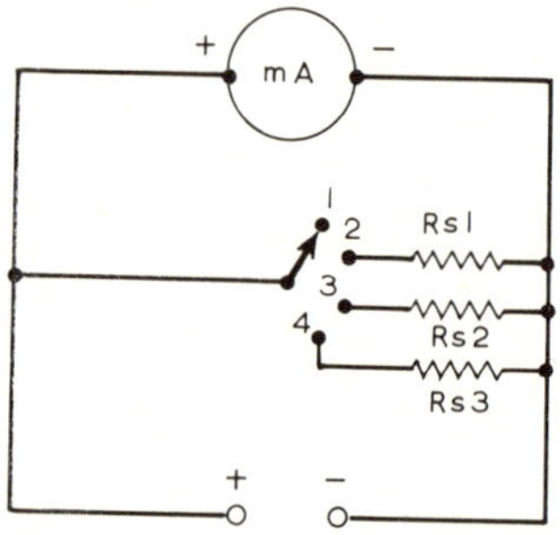

FIG. 2.4. *Switched shunts. This simple method is liable to introduce errors due to contact resistance at the switch, and is not used in commercial instruments.*

and this pulled up and twisted together as a loop, as shown in Fig. 2.3. Very fine resistance adjustment can then be made by twisting together more of the wire if the reading is too high or by untwisting if the reading is too low. To facilitate the measurement of current over several ranges, a series of shunts can be arranged in a switched circuit, as shown in Fig. 2.4.

CURRENT TESTING

Current testing is usually quite straightforward, since the meter is always connected in series with the various components in the circuit. For example, if it is required to check the total h.t. current taken by a battery radio set, the meter is connected in series with either the h.t. positive or h.t. negative lead to the battery; ensure that the meter is connected with correct polarity (see Fig. 2.5). The same applies when making an l.t. current check in battery receivers.

In mains receivers, the h.t. feed should be broken and the meter connected to restore the circuit, as shown in Fig. 2.6. Current can be checked in any circuit by breaking the circuit and introducing the meter. If the current in the circuit is far in excess of the full-scale deflection of the meter, the meter will almost

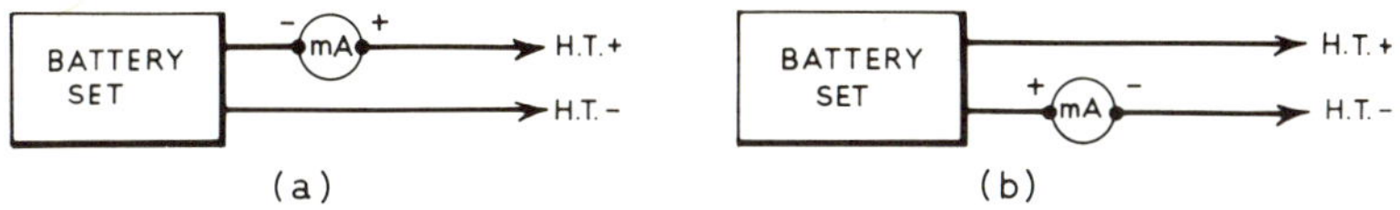

FIG. 2.5. *Method of measuring total h.t. current in a battery receiver:* (a) *with meter in h.t. positive lead, and* (b) *with meter in h.t. negative lead. Note the polarity of the meter in each case.*

certainly burn out, so one should always be cautious and set the meter to a high current range before switching on the receiver or circuit under test. The small reading on the high range will provide an indication as to the range that is required in order to give a more accurate measurement. Never break a circuit and attempt to insert a meter with the equipment switched on, and never switch the meter from one current range to another without first switching off the test circuit.

The meter should always be inserted in the circuit at a point of low or zero signal potential. For example, if it is required to measure the current taken by the anode circuit of an intermediate-frequency amplifier, the meter should be inserted between the primary of the i.f. transformer and the h.t. line, and *not*

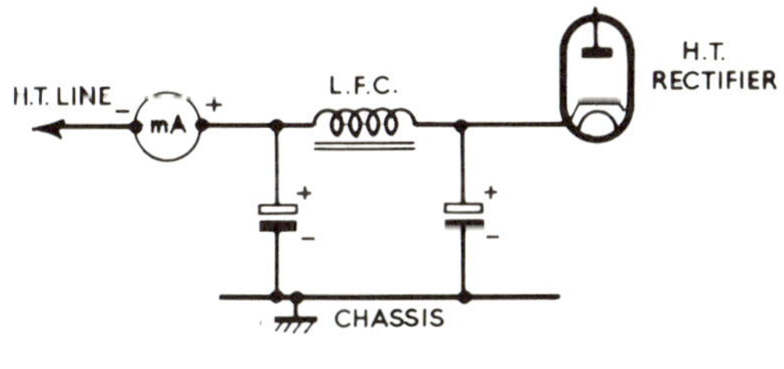

FIG. 2.6. *Method of measuring total h.t. current in a mains receiver. The h.t. feed should be broken and the meter inserted at the point shown.*

between the anode and the winding (see Fig. 2.7*a*). The reason for this is that signal currents as well as anode current are present at the anode, and while the meter will only indicate the direct current, the meter and connecting leads may disturb the normal operation of the circuit and in this way give an incorrect reading. Moreover, meter decoupling may be required in certain cases to avoid instability or feedback effects, due to the meter and connecting leads, when measuring current in high-gain circuits or in circuits which do not have a true zero signal potential point at which to introduce the meter (Fig. 2.7*b*). The same rules apply when testing in transistor equipment.

Current testing can provide useful information concerning the condition of various components in a receiver. For instance, low anode current may indicate

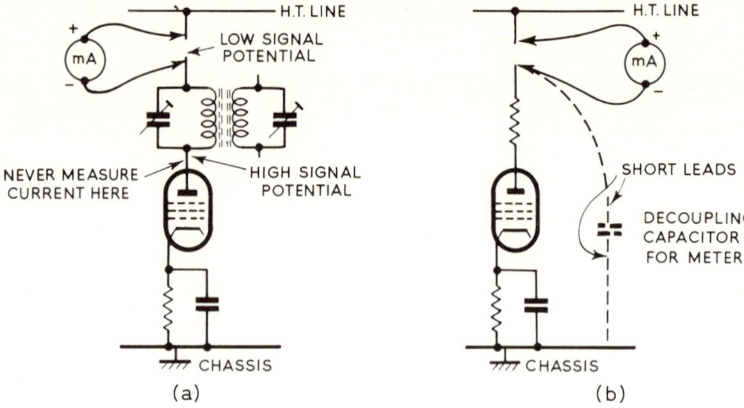

FIG. 2.7. *When checking current, the meter should always be inserted in the circuit at the point of low or zero potential, as shown at* (a), *which illustrates the method of measuring the anode current of an i.f. amplifier.* (b) *In certain cases, when measuring current in high-gain circuits, decoupling may be required on the meter to avoid feedback and false readings.*

that a valve or valves are losing emission or, in a battery set, that the h.t. battery is in need of replacement. In battery sets, also, low h.t. current may indicate that the l.t. battery is partially exhausted. This can be proved by following up with an l.t. current check.

Lack of h.t. current may mean that the h.t. battery or h.t. rectifier is dead. It may also mean that the on/off switch is faulty or that a fuse has blown; in modern low-consumption portable receivers, this symptom may be caused by failure of the filaments of all the valves owing to incorrect testing or a short-circuit.

The current taken by the anode circuits in ordinary receivers is relatively low, in the region of milliamperes (e.g. 0·001 ampere), and in some cases less than 1 mA. For instance, the current in the grid resistor of a local oscillator may be around 50 microamperes (a microampere, μA, is one millionth of an ampere or one thousandth of a milliampere). Thus, very sensitive meters are required for

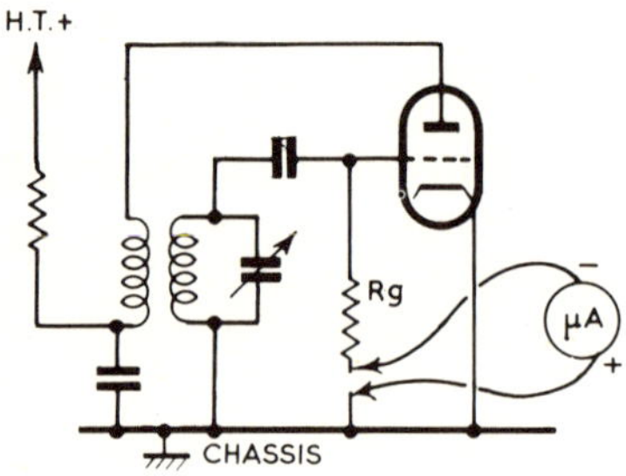

FIG. 2.8. *The condition of an oscillator may be checked by measuring the grid current. When the valve is not oscillating, the current will be negligible, in the region of 10 micro-amperes, but will be much higher when the valve is oscillating. It is normal for the grid current to vary when a receiver is tuned over a waveband.*

this kind of testing. However, if such a meter is available a check of the condition of the local oscillator in any receiver can be made by introducing the meter in series with the oscillator grid resistor, as shown in Fig. 2.8. If the oscillator is not working, there will be negligible grid current.

APPLICATION OF D.C. METERS

VOLTAGE MEASUREMENT

Voltage is measured on a moving-coil meter by first determining the current through the meter resulting from the potential developed across the meter resistance and an external resistance, known as a multiplier. The scale is then calibrated in voltage instead of current.

As the internal resistance of most meters is relatively small, on all but possibly the lowest voltage range a multiplier will need to be used, as shown in Fig. 2.9. The value of the multiplier may be obtained from the following expression:

$$R_v = (E \times 10^3/I) - R_m$$

where R_v is the multiplier in ohms; E is the required full-scale voltage to be measured in volts; R_m is the internal resistance of the meter; and I the full-scale current of the meter in milliamperes. For high voltage ranges, the internal resistance of the meter can usually be ignored.

As an example, supposing it is required to arrange a 1-mA meter having an internal resistance of 500 ohms to measure 10 volts full-scale. The expression thus reads: $R_v = 10 \times 1{,}000/1 - 500 = 9{,}500$ ohms.

The accuracy of a voltmeter is governed by the accuracy of the multiplier as well as by the accuracy of the meter movement itself. High stability resistors are

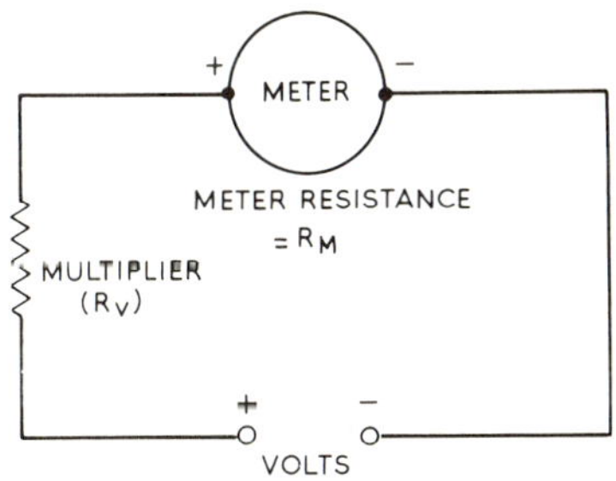

FIG. 2.9. *A simple multiplier connected in series with a current meter for the measurement of voltage.*

readily obtainable commercially, and these make good multipliers. The multiplier must also be capable of sustaining the required voltage drop across its body. High-voltage resistors are available, but most ordinary resistors are unable to stand more than about 1,000 volts without flashover or breakdown, and for this reason high-voltage multipliers are usually formed of a series of resistors whose total value equals R_v.

Meter Sensitivity

If a voltmeter is designed round, say, a 1-mA meter movement and the full-scale deflection is 100 volts, it is obvious that the test circuit will have to supply the meter circuit with 1 mA of current in order to provide full-scale deflection on the meter. Although a potential of 100 volts may exist across the test circuit, there may not be sufficient current available in the circuit to work the meter. When this happens, the test circuit is loaded by the meter circuit; the test circuit potential falls and the meter reads something less than the potential present when the meter is not connected.

The more sensitive the meter movement, the less the current required from the test circuit and the more accurate will be the reading. Meter movement sensitivity is expressed in terms of the current required for full-scale deflection. However, with voltmeters, the sensitivity is often given in terms of ohms-per-volt. The ohms part is the total resistance of the meter and the multiplier necessary to read 1 volt full-scale deflection. This is simple Ohm's Law, where the resistance R is equal to the full-scale voltage E divided by the full-scale current of the meter I. Thus, the total resistance of a voltmeter (including the internal resistance of the movement) to read 1 volt full-scale on, say, a 1-mA movement is 1,000 ohms, which means that such a movement in a voltmeter would have a sensitivity of 1,000 ohms-per-volt (e.g. $R = 1/0{\cdot}001 = 1{,}000$).

Similarly, a 2-mA movement will give a sensitivity of 500 ohms-per-volt. Higher sensitivities of the order of 10,000 and 20,000 ohms-per-volt require movements of 100 microamperes and 50 microamperes full-scale. Super-sensitivity, such as 100,000 ohms-per-volt, demands a movement of only 10 microamperes full-scale.

METER LOADING ON TEST CIRCUITS

When a voltmeter is connected across a test circuit, the effect on the circuit is equivalent to a resistor being connected. The value of the resistor is governed by the sensitivity of the instrument and by the range to which it is set. For example, a meter with a sensitivity of 1,000 ohms-per-volt, set to 1 volt full-scale, will shunt the test circuit by 1,000 ohms. However, the same meter set to 100 volts full-scale will shunt the circuit by 100,000 ohms. A meter with a far higher sensitivity—say, 100,000 ohms-per-volt—will, on similar ranges, shunt the test circuit by 100,000 ohms and 10 megohms respectively.

In Fig. 2.10*a* is shown a simple potential divider comprising two resistors in series connected across a battery. Provided the voltage of the battery and the value of the resistors are known, the actual voltage across either resistor can be worked out without any difficulty by the following expressions:

$$\text{Voltage across } R_1(E_{R1}) = \frac{R_1 \times \text{Battery voltage } (E)}{R_1 + R_2}$$

$$\text{Voltage across } R_2(E_{R2}) = \frac{R_2 \times \text{Battery voltage } (E)}{R_1 + R_2}$$

Thus, if R_1 and R_2 are like values, then the voltage across each resistor will be half the battery voltage. However, although this is the actual voltage when the circuit is working normally, as soon as a voltmeter is connected across one of the resistors to measure the voltage, the circuit conditions are altered because the voltmeter itself then becomes part of the circuit.

This is shown in Fig. 2.10*b*. Here R_1 and R_2 are 100,000 ohms, E is 100 volts and the voltmeter is 1,000 ohms-per-volt, set to 50 volts full-scale. The resistance of the voltmeter is 50,000 ohms, which is in parallel with R_2. The total resistance (R_t) of R_2 and the meter in parallel is equal to $R_m \times R_2/R_m+R_2$, where R_m is the meter resistance on the 50-volt range. This works out to 33,333 ohms. The circuit conditions are now as shown in Fig. 2.10*c*, and the voltage across R_t is only a little over 24 volts. This will be the indication given by the voltmeter, which is greater than 50 per cent below the true voltage when the meter is disconnected.

The error will not be so great when a voltmeter of higher sensitivity than 1,000 ohms-per-volt is used to measure the voltage, but there will, nevertheless,

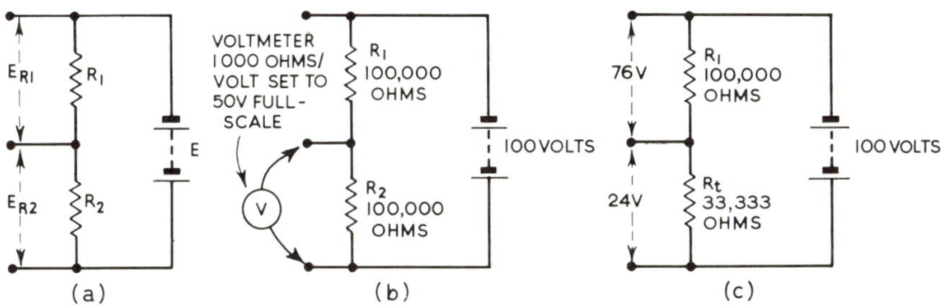

FIG. 2.10. (a) *Simple potential divider, whose conditions are altered by the connexion of a voltmeter, as at* (b), *to those shown at* (c).

be an error. True voltage in high-resistance circuits can never be measured accurately, and since most voltage measurements in radio circuits are made across relatively high resistances, this should always be borne in mind.

Other things also happen in the circuit when a voltmeter is connected. As seen in Fig. 2.10*c* for example, the voltage across R_1 rises in proportion to the fall across R_2 when the voltmeter is connected. Moreover, the total current from the battery also increases when the voltmeter is connected. This increase represents the current required to deflect the meter's pointer over the scale. In a piece of electronic circuit, the source of power may not be a battery, but another high-resistance circuit, and this may mean that the meter loading will be reflected into other sections of the circuit.

Beginners are often puzzled as to why a small deflection on a voltmeter when making a circuit test does not increase substantially when the voltmeter is switched to a lower voltage range. This, of course, is due to the increased loading of the voltmeter holding the circuit more or less in balance from the current point of view. If a test circuit is capable of supplying only a limited external current to operate the meter, then the limiting factor is in the circuit and not in the meter, so from the aspect of pointer deflection this will remain relatively constant irrespective of the resistance value of the multiplier switched in. However, the most accurate reading will be that given on the highest voltage range.

As most voltage measurements are made in valve or transistor circuits, the simple potential-divider illustration of meter loading is often varied in practice.

For example, in Fig. 2.11 is shown the basic circuit of an a.f. amplifier, in which it is desired to measure the voltage on the screen of the valve. It is probable that a voltmeter with a sensitivity of 1,000 ohms-per-volt and 100-volt full-scale would be employed, making a total of 100,000 ohms. A typical value for the screen resistor (R_s) in such a circuit is 500,000 ohms. With a h.t. voltage of 200 volts, as shown, and assuming that the screen of the valve draws 0·2 mA, the drop across R_s would be 100 volts, thus producing a potential of 100 volts between the screen and receiver chassis. This is representative of the actual,

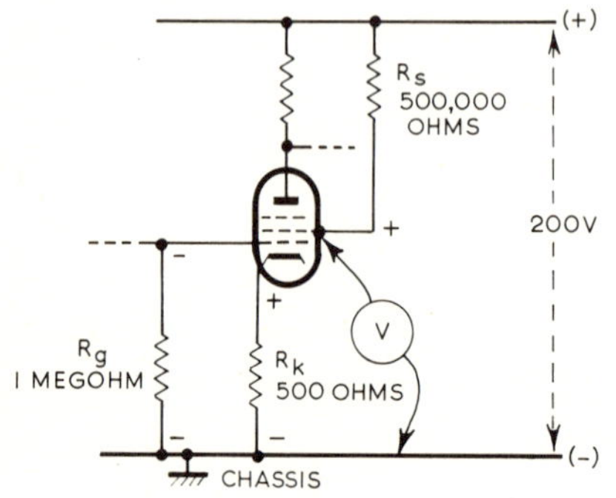

FIG. 2.11. *The error introduced by the connexion of a voltmeter in the position shown is described in the text.*

undisturbed operating condition of the screen. If the screen-to-cathode path of the valve is looked upon as a pure resistance, whose value would be 500,000 ohms, then an error based upon the analysis of Fig. 2.10 would be caused by the connexion of the voltmeter.

However, this is not strictly true, since the reduced voltage at the screen will reduce the screen current, for it is unlikely to continue taking 0·2 mA when the voltage drops from 100 to around 30. In effect, this will reduce the voltage drop across the screen feed resistor and consequently raise the screen voltage, but not to the same extent that the loading of the voltmeter reduces it.

To summarise, it has been shown that the connexion of a voltmeter will so affect the test circuit potential that an incorrect reading is obtained, to which it is often extremely difficult to apply a correction factor. This happens when the meter is connected across any two points where a high resistance is introduced in series with the meter; that is, a resistance which is high in comparison with the internal resistance of the meter itself.

The error can be neglected when the meter is connected in low-resistance circuits. For example, the true e.m.f. of a battery will almost invariably be given by connecting a voltmeter across its terminals. This is because the internal resistance of a battery should be very low in comparison with the meter resistance, though this may not be so when the battery is partially exhausted and itself has a high internal resistance. Similarly, the voltage measured across the cathode resistance (R_k) of the circuit in Fig. 2.11 would be almost the actual voltage, since R_k is usually relatively small.

The voltage drop across R_k provides grid bias for the valve, since the resistor causes the valve cathode to become positive with respect to the control grid of the valve, which is returned to chassis (the h.t. negative line) through the grid resistor (R_g). However, one would not expect to secure a correct grid-bias voltage reading by connecting the voltmeter between the valve cathode and grid (negative to grid), as R_g is usually of high value, in the region of one megohm (1,000,000 ohms) or so, which would be in series with the voltmeter. The grid-bias voltage is, therefore, always measured between cathode and the point to which the grid resistor is returned, which is usually the receiver chassis.

Testing in transistor circuits calls for low-scaled voltage ranges and movements of high sensitivity for measuring low current values. Moreover, to avoid the loading of the testmeter from significantly changing the operating parameters of the transistor circuit under test, the d.c. meter should desirably have a value of about 20,000 ohms-per-volt or more.

Fig. 2.12 shows a simple transistor amplifier circuit in which voltage tests often have to be performed. Six such tests are, in fact, indicated on the diagram.

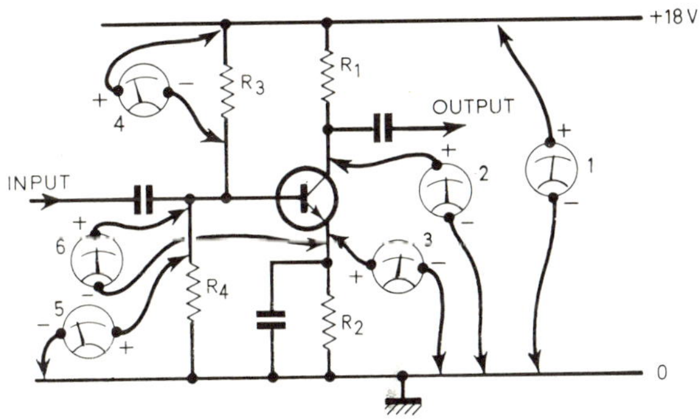

FIG. 2.12. *Showing the points of voltage measurement in a simple transistor circuit. These are fully described in the text.*

The transistor is an n-p-n type which means that the collector is connected to a positive potential relative to the emitter and that the base is slightly positive with respect to the emitter to give forward current in the emitter/base junction. Under these conditions the transistor conducts and current flows in the emitter/collector circuit.

To check the d.c. conditions in and around this kind of circuit one would probably first measure the supply line potential. This is shown by test 1, and for this the voltmeter should have a full-scale deflection of about 20 or 25 V (e.g., just a little greater than the supply line potential). Meter sensitivity is not important for this application unless the supply line is fed from a high resistance source, which is not very likely. Usually, the supply source resistance is in the hundreds of ohms (sometimes even lower), possibly rising to the thousands of ohms, but even this is not a high resistance relative to the meter resistance, so the volts drop across it due to a meter current as high as 0·5 mA would give insignificant readout error.

Test 2 shows a measurement of collector voltage. This time the current for the meter flows through the collector resistor R_1, and if this is relatively high some error in measurement could occur, just as it can when measuring the anode voltage of a valve via the anode load resistor. However, transistor circuits commonly employ resistors of considerably lower value than valve counterparts, 4·7 k being a typical value for a collector load in the type of circuit shown in Fig. 2.12. If it is assumed that the transistor is conducting, then the collector voltage will be something less than the supply line voltage by the amount of the voltage dropped across the load resistor. Again, then, a meter of mediocre sensitivity could possibly be adopted for this measurement, but remember the full-scale deflection should not be much more than 20 V, less if substantial current is flowing through the load, based on a line voltage of 18.

Since the transistor is conducting, the transistor and the emitter resistor R_2 can be considered as the bottom leg of a potential divider, with R_1 as the top leg, at which point the voltmeter is connected. The meter is thus effectively measuring the volts drop across the transistor and R_2 in series. If the transistor is switched 'hard on', most voltage will be dropped across the emitter resistor, and this can be measured by test 3. In amplifier configurations, however, the transistor idling current is relatively small, the transistor being biased to class A working. We can easily calculate the current by measuring the voltage across the emitter resistor R_2 (test 3) and dividing it by the value of R_2. If R_2 is given in kilohms the answer will be in mA. For example, 1 V measured across 1 k implies a current of 1 mA. This is really the best way of determining current in a transistor circuit since it avoids having to make resistor disconnections from the printed circuit board.

The accuracy, of course, will depend on the resistive shunt effect of the meter, but since the emitter resistor is usually of fairly low value the error should not be large provided the meter is not much less than, say, 10,000 ohms-per-volt. We must remember, of course, that for this measurement a full-scale deflection of, perhaps, 1 to 1·5 V is necessary for accuracy. A meter of a mere 1,000 ohms-per-volt on the 1 V range, of course, would shunt the emitter resistor by 1 k, which certainly would produce an error.

Resistors R_3 and R_4 form a potential divider across the supply which feeds the base at their junction. It sometimes becomes necessary to measure the voltage across R_3 and R_4, as shown by tests 4 and 5. The normal through current of the potential divider is modified by the base circuit pulling current from it, and the amount of current taken by the base depends on the voltage (relative to emitter) applied to it. Thus we have to be careful that meter shunting effects do not significantly change the ratio of the potential divider, and bearing in mind the small voltage involved the meter has to be operated on the lowest voltage range. If the meter of test 4 causes a fairly heavy shunt across R_3 the transistor will be turned on harder and under certain conditions damage could result.

Probably the best way of measuring base voltage is by test 6. Here the meter is connected directly across the base and emitter and shunting would merely turn down the transistor current. Nevertheless, a highly sensitive meter is required for this test because the base/emitter voltage may be only a fraction of 1 V. For this test an electronic testmeter comes into its own, especially if it has a full-scale deflection of 500 mV or less. However, special care must be taken when measuring voltages in the base circuit with an electronic testmeter powered from the mains supply, for intermittent and faulty connections can incite transient currents of a magnitude sometimes sufficient to damage the emitter/base junction of small-signal transistors. Field-effect transistors are particularly vulnerable in this respect.

Quite a lot of information can be gleaned by d.c. testing in transistor circuits provided the approach is logical. We can, in fact determine whether or not the transistor is actually working from the d.c. point of view at least by, for example, monitoring the voltage across the emitter resistor while changing the base bias. It is always desirable to cause a reduction in base bias rather than an increase for the reason already mentioned, and the best way of doing this is to shunt the lower arm of the base potential divider with a resistor of equal or lower value than that of the arm itself. If the transistor is working the emitter/collector current will fall, as also will the voltage dropped by the emitter resistor.

Shorting or open-circuit transistor junctions can often be revealed in this way without having to extract the device from the printed circuit board, which is rarely a very happy exercise!

To summarise, therefore, although in general transistor circuits employ resistive networks of significantly lower values than in valve counterparts, the voltages of the former are often much smaller than of the latter. Transistor testing thus calls for a testmeter with a low voltage full-scale deflection range, and because this means that the value of the series resistance is reduced the testmeter must possess an intrinsically high sensitivity.

RESISTANCE MEASUREMENT

The basis of the simple ohmmeter is a moving-coil current meter. This requires the use of a battery which, when set up in an ohmmeter circuit, measures the current which a known voltage will drive through an unknown resistance. To obviate the need for a voltmeter to check the battery voltage, one or two variable resistors are also connected in the circuit, and are so adjusted that the pointer rests at 'zero ohms' or full-scale when the test terminals are short-circuited.

In Fig. 2.13*a* is shown the basic circuit often used for the measurement of high and medium resistance values. The circuit is first set up by short-circuiting the two terminals and bringing the meter to full-scale deflection by the variable resistor R_v. The short-circuit is then removed from the terminals and the unknown resistor connected across them. This will cause a smaller current to flow in the meter circuit, and the pointer will indicate something less than the full-scale value. By comparing this new reading with the full-scale current of the meter, the unknown resistance value can be calculated from the following expression:

$$R_x = (R_t \times i)/I - R_t$$

where R_x is the unknown resistance in ohms; R_t is the total resistance of the ohmmeter circuit, including R_f, R_v, the internal resistance of the meter movement and the internal resistance of the battery; i is the full-scale current of the meter; and I is the new current reading when the unknown resistance is connected in circuit.

From Fig. 2.13*a* it will be seen that R_f and R_v in series with the meter movement represents, in fact, a voltmeter which is adjusted to provide full-scale deflection from the voltage given by the battery. The variable resistor is included to allow the movement to be brought to full-scale even though the battery voltage may be slightly less than it should be, owing to partial exhaustion. If it is found impossible to 'zero' the meter after prolonged use, the battery should be replaced.

In Fig. 2.13*b* is shown an alternative ohmmeter circuit for low-value resistances. This is a parallel-connected arrangement, as distinct from the series-connected arrangement in Fig. 2.13*a*.

With the parallel circuit, the movement is first brought to full-scale deflection by adjusting R_v with the terminals *disconnected*. When the unknown resistance is connected, the current indication will fall (since the unknown then acts as a meter shunt), and the resistance value may be found from the following expression:

$$R_x = (R_m \times I)/(i - I)$$

where, in this case, R_m is the internal resistance of the meter movement only and the other factors are as in the previous expression.

An arrangement which is somewhat more accurate than the two circuits just described is shown in Fig. 2.13*c*. This is, in effect, an ordinary voltmeter circuit,

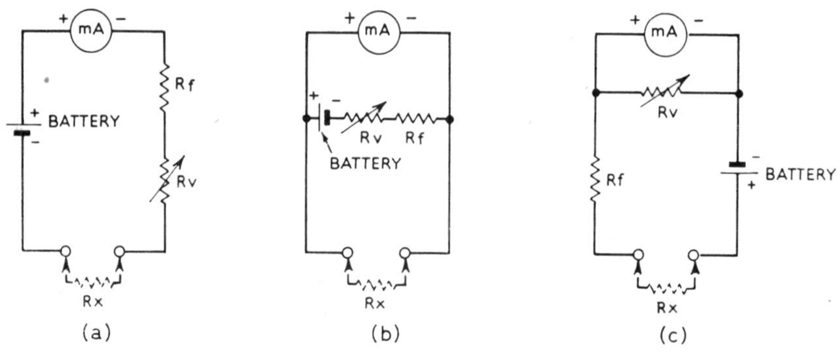

FIG. 2.13. (a) *Series-connected ohmmeter circuit for measuring medium and high values of resistance.* (b) *Parallel-connected circuit for measuring low resistance values.* (c) *Ohmmeter with variable shunt.*

with R_f as the multiplier, but in addition has a variable shunt R_v, which is the 'set zero' control, and is adjusted to give full-scale indication on the meter movement when the terminals are shorted. The unknown resistance value R_x may be calculated from the following expression:

$$R_x = R_f(i/I - 1)$$

where R_f is the series resistor; i is the full-scale current of the meter (e.g.,

i = battery voltage/R_f); and I is the new current reading when R_x is connected (e.g., I = battery voltage/$R_f + R_x$).

Resistance meters and 'ohms' scales of multirange meters are calibrated directly in terms of resistance, but the normal calibration of current meters could be employed in association with an 'ohms' conversion chart.

Resistance may, of course, be measured by other means, the best known of which is the Wheatstone bridge, which is dealt with in Chapter 7. The portable ohmmeter is, nevertheless, the most convenient and practical instrument for day-to-day fault finding and for continuity and resistance checks in electronic equipment.

DIGITAL VOLTMETER

The input amplifier of most digital voltmeters is designed to have a high input impedance. Such instruments, therefore, are ideal for testing in high resistance circuits, and in this respect are akin to the valve or transistor voltmeter. As already noted elsewhere, the digital instrument is more costly than an analogue instrument endowed with similar facilities. In fact at the time of writing the lowest-priced digital multimeter is several times that of an equivalent analogue multimeter. Nevertheless, the signs are that the digital system will eventually achieve price parity with the analogue meter.

Although the conventional analogue multimeter is a very potent servicing type of instrument, combining simplicity of operation with wide-range accuracy and portability, the readout of the digital instrument yields not only increased resolution and accuracy but it also significantly reduces operator fatigue and virtually deletes reading error. At the present time, however, it is finding greatest application in research and development laboratories and for the servicing of computer equipment.

ELECTROSTATIC VOLTMETER

The electrostatic voltmeter does not rely on current to move its pointer over the scale. It is a meter in which deflection is accomplished by an electrostatic attraction or repulsion between two plates when the potential to be measured is applied across them. One plate is fixed and the other secured to the pivot carrying the pointer. As with other instruments, very fine hairsprings are employed to zero the movement when at rest.

Although the electrostatic voltmeter can be arranged to measure alternating voltage as well as direct voltage, it is dealt with here because from the servicing point of view it is mainly used for the measurement of e.h.t. voltages in television receivers. On d.c., apart from the initial charging kick, which is negligible, it imposes no load at all on the circuit under test. However, from the a.c. aspect, it appears as a capacitance, and in high-frequency circuits is sometimes just as objectional for this reason as are other forms of meters with relatively low impedance.

Electrostatic voltmeters are intrinsically single-range instruments, for in order to extend their range on d.c. a resistive potential-divider is required which, of course, consumes power from the test circuit and thus defeats the whole object of the instrument. On a.c., their range can be extended by a capacitive potential-divider, but, again, this means that the instrument will impose an even greater impedance loading on a high-frequency circuit.

The scales are basically square-law, and are usually very cramped at the bottom end.

E.H.T. MEASUREMENTS

The electrostatic voltmeter was at one time extensively employed for the measurement of e.h.t. voltages in television sets. However, with the advent of the sensitive moving-coil movement range extension by means of a high-voltage probe (see Fig. 1.15) is becoming more commonplace. The high series resistance required to extend the range of, say, a 20,000 ohms-per-volt meter to read 30 kV full-scale puts very little load on the e.h.t. voltage source and thus quite reasonable readout accuracy is achieved without distressing the e.h.t. generator. The e.h.t. high-voltage probe contains a chain of resistors of suitable voltage rating to provide the required total series resistance for the intrinsic sensitivity of the instrument with which it is to be partnered.

Colour television sets, in particular, need to be adjusted for a rather critical e.h.t. voltage, since excessive voltage can yield x-ray emission beyond the safe count while inadequate voltage can cause a poor colour display of low average brightness. To avoid the service technician having to make absolute e.h.t. voltage measurements, many colour television sets incorporate a test point at which the voltage developed across a resistor of specific value in series with the e.h.t. supply can be measured. Thus by knowing the value of this resistor, the voltage read across it can be apportioned to the e.h.t. voltage actually present on the final anodes of the picture tube (see my *Colour Television Servicing*, by the same publisher).

CHAPTER 3

Application of A.C. Meters

ALTHOUGH there are several types of instruments which respond to alternating current, the moving-coil instrument responds only to direct current. In order to measure a.c. with a moving-coil meter, therefore, it is necessary first to rectify the alternating current and then apply the resultant d.c. to the meter movement. Full-wave rectification is essential for this operation, and to provide the maximum conversion of alternating current to direct current four rectifier elements arranged in the form of a bridge circuit are employed.

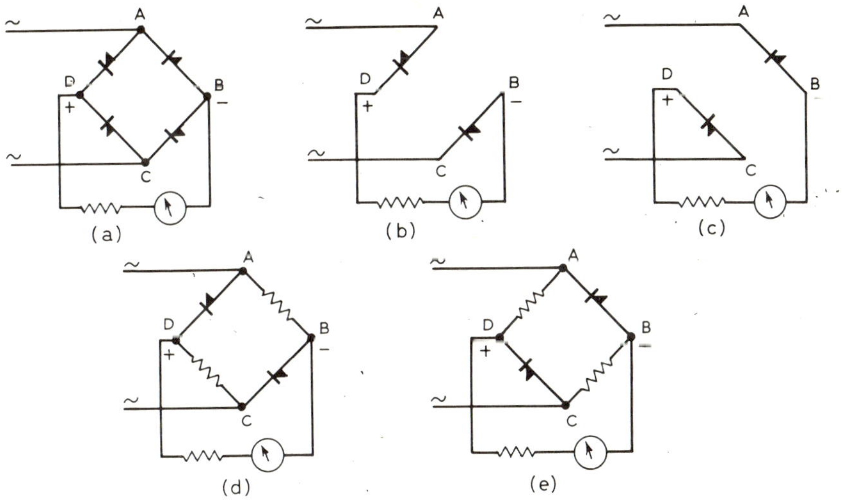

FIG. 3.1. *Operation of an instrument bridge rectifier.* (a) *The complete circuit, including the meter.* (b) *and* (c) *Path of the current during the first and second half-cycles of alternating current.* (d) *and* (e) *Showing the effective leakage paths of those elements not working in a forward direction.*

Fig. 3.1 illustrates the operation of the bridge rectifier. The rectifier symbol indicates the conventional flow of current, and diagram (*a*) shows the full basic circuit of the bridge rectifier and meter. Diagram (*b*) illustrates the operation of the rectifier in one half-cycle. In a similar manner, diagram (*c*) shows the operation on the second half-cycle, when the current is flowing in the opposite direction. It will be seen, therefore, that current flows continuously in the same direction through the moving-coil, although it has reversed in the external parts of the circuit.

This function of the rectifier is well known, but it must be remembered that no form of 'semi-conductor' rectifier is perfect, for at its best it will show a low forward resistance and a high reverse or leakage resistance. It will thus be appreciated that the meter is at an apparent mid-potential between the alternating input. In diagram (*d*), which is a more exact representation of diagram (*b*), the elements *D*, *C* and *A*, *B* are not functioning in a forward direction for the particular half-cycle shown, but act as high-resistance 'leaks'.

Current entering at *A* passes to *B* mainly through the element *A*, *D* and the moving-coil, but a small amount by-passes the moving coil and arrives at *B* by leakage through element *A*, *B* (which ideally should be a perfect insulator). Moreover, all the current which passes through *A*, *D* does not traverse the moving-coil since it can reach *C* through the element *D*, *C*, thus by-passing the member and the element *B*, *C*. The result is that the current which passes through the moving-coil is reduced from the total available by the sum of the two 'leaks'. In a corresponding manner, the reverse cycle is shown in diagram (*e*) which is a more exact representation of diagram (*c*).

There are many materials in the 'semi-conductor' category, such as copper oxide, germanium, silicon and selenium, which can be used for instrument rectifiers, but copper-oxide rectifiers are often favoured, as this material exhibits a useful characteristic for certain forms of rectifier moving-coil instruments.

Under normal conditions, the rectification system works adequately without wear or tear for long periods, but if an instrument is overloaded, the voltage across the rectifier input may rise to a dangerous level. A condition may arise where the voltage across the meter and one operative rectifier element is sufficiently great to puncture the element which is acting as a 'leak' across them. The higher the resistance of the meter, the greater will be the rise in voltage across those elements acting as 'leaks', when the meter is overloaded. This shortcoming is characteristic of all high-sensitivity instruments, which should always be treated carefully when they are used to measure alternating current.

Rectifier-type instruments can also be easily damaged by voltage transients; these are very steep-fronted current pulses which may be created by the opening and closing of switches, charging and discharging of capacitors and so on. Some transients may persist only for a few milliseconds and are unreadable on normal moving-coil instruments, but their duration may be long enough to puncture the rectifier elements. It is, of course, practically impossible to avoid the occurrence of surges, but if their damaging effect is known, steps can be taken to ensure that the minimum number of switching operations is carried out with the a.c. meter connected in circuit. Certain instruments are fitted with a surge-suppression rectifier to provide an additional measure of protection.

A.C. SENSITIVITY

The sensitivity of the meter on a.c. ranges is a function not only of the meter movement but also of the rectifier. High-sensitivity rectifiers, having low leakage currents, are available, but unfortunately some early ones tended to lose this characteristic rapidly with overload and gradually with age. Recent ones are less prone to this fault.

In the case of a meter movement which requires 1 mA to bring it to full-scale, a few microamperes of rectifier leakage current is of no consequence, for the instrument will continue to have a high degree of accuracy even should the rectifier leakage increase during service, as is often the case. On the other hand, if a 50-microampere full-scale movement is used instead, a leakage of a few microamperes becomes disastrous for, while correction for the leakage can be made at full-scale deflection when the instrument is new, compensation cannot be applied down the scale owing to the non-linearity of the rectifier characteristics. Even slightly increased leakage current due to age will cause the instrument to give widely erroneous readings.

It is for this reason that the sensitivity of multirange instruments is often less on the a.c. ranges than on the d.c. ranges. This ensures that the instrument will continue to read correctly on the a.c. ranges, and be less delicate from the servicing viewpoint, over the maximum length of time. High-sensitivity a.c. meters are available, however.

The scales of a.c. rectifier instruments become slightly cramped towards the low-current end. This is because the resistance of the rectifiers decreases as the current through them increases. However, on some instruments this factor may be compensated, in which case both d.c. and a.c. scales are linear.

A.C. MEASUREMENT

Fig. 3.2 shows a simplified circuit of the a.c. ranges of an Avo Model 8. As is usual practice, the multiplier elements, for the measurement of voltage, are connected between one of the a.c. terminals and the rectifier, so as to avoid the rectifier receiving the full measured a.c. voltage which would undoubtedly destroy it. Transformer TR_1 is tapped and works as an auto-transformer on the current ranges and the low-voltage ranges. Resistors R_2 to R_3 are multipliers.

The current to be measured is passed through the appropriate windings of the transformer, and the voltage induced into the other windings is applied directly across the bridge rectifier, MR_1. The d.c. voltage from the output of the rectifier is applied to the meter, via the shunt 'potential divider' arrangement of resistors.

The 'swamp' resistor, which is a feature of most multirange instruments, is included to build up the complete internal resistance of the movement. In the event of the instrument being damaged, it is possible that its internal resistance would be altered as the result of the repair. The swamp resistor enables this to be compensated for by decreasing or increasing its value, so that the combined resistance remains unchanged. Without this, an alteration in meter internal resistance would demand alteration to most of the shunts and multipliers which, of course, would prove costly. The swamp resistance may sometimes serve to compensate for the effect of temperature variation.

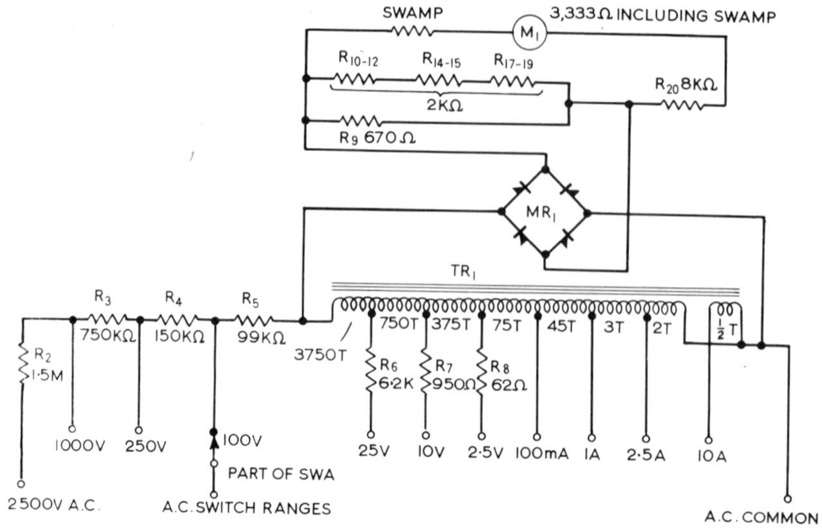

FIG. 3.2. *Simplified circuit of the a.c. ranges of a Model 8 Avo. The tapped transformer* TR_1 *works as an auto-transformer on the current ranges and the low-voltage ranges.*

The current supplied by the bridge rectifier to the meter coil causes the movement to deflect in accordance with the average value of the current, which is equal to 0·636 of the peak value. However, the instrument scale is usually calibrated in terms of the root-mean-square (r.m.s.) value of the applied alternating potential or current, which is equal to 0·707 of the peak value. The ratio of the average value to the r.m.s. value of the voltage or current is thus approximately 1·11 (i.e., 0·707/0·636). This means that the form-factor of the applied voltage or current will influence the reading to some degree, for the scale calibration is usually made by employing sine-wave potentials. If the waveform is very square, with a flat top, then the reading will be high, since it is then approaching direct-potential conditions, which would cause the instrument to read about 11 per cent high. On the other hand, if the waveform is of a 'peaky' nature the reading will be low.

The frequency of the voltage or current applied to a rectifier-type instrument also influences the reading. This is because the rectifier elements have capacitance which tends to by-pass more and more of the current from the moving-coil with increase in frequency. The frequency characteristics of the current transformer, such as TR_1 in Fig. 3.2, also have a bearing on this. Nevertheless, most rectifier-type instruments are useful up to at least 10,000 Hz, which takes in most of the audio frequencies, and thus enables the instrument to be used for the measurement of audio voltages and a.f. output levels.

Some a.c. meters are specially designed and calibrated (in watts, voltage and decibels) for audio-frequency applications. These, however, usually have a frequency response which is superior to that of the ordinary a.c. voltmeter designed essentially for the measurement of 50 Hz or 60 Hz voltages of sine-wave form.

APPLICATION OF A.C. METERS

THERMOCOUPLE METER

This type of meter is suitable for the measurement of a.c. from very low frequencies up to very high radio frequencies, and is often used to measure the r.f. aerial current in transmitters. The meter proper consists of an ordinary moving-coil movement, whose deflection relies upon the direct current produced by a thermocouple when this is heated by the applied alternating current. The thermocouple comprises a junction made up of two dissimilar metals which have the property of generating a small current when heated. The thermo-electric effect is measured in terms of microvolts per degree rise in temperature, and thermocouples possessing a relatively high 'thermo-electric power' are antimony with platinum, and copper partnered with constantan. There are, however, other combinations which are sometimes used.

The e.m.f. produced at the junction of the thermocouple is proportional to the applied heat (and also depends on the metals forming the junction), and since the heating effect is proportional to the square of the current multiplied by the resistance (i.e., $I^2 \times R$), the meter deflection follows a square law. That is, the scale is considerably non-linear, and closes up towards the low-current end. Such meters indicate true r.m.s. values irrespective of the form-factor of the applied current, because they are operated by heating effects which themselves are based on r.m.s. values. Direct current applied to the meter will, of course, also be measured accurately for the same reason.

MULTIMETER

Most of the basic principles described in previous chapters are applied in the popular multirange meter, often called the 'multimeter'. There are two versions

FIG. 3.3 *Model 8 AvoMeter, a 'universal' type suitable for use on a.c. and d.c. circuits. This popular multimeter has 30 ranges.*

FIG. 3.4. *Internal construction of the Model 8 AvoMeter.*

in general use: the d.c.-only type, and the a.c./d.c. or 'universal' type. An instrument of the latter type is illustrated in Fig. 3.3; the interior is shown in Fig. 3.4.

The complexity of this kind of instrument depends on the number of ranges it provides. The circuit in Fig. 3.5 is that of the Avo Model 8, which has 30 ranges. A simplified circuit of the a.c. ranges of this instrument is given in Fig. 3.2. Other simplified circuits of the d.c. ranges, the ohms range, the 'ohms × 100' range, the 'ohms ÷ 100' range and the 'insulation' range are given in Figs. 3.6, 3.7, 3.8, 3.9 and 3.10 respectively.

MEASUREMENTS AT LOW FREQUENCIES

The rectifier moving-coil instrument, either as an individual instrument or as a multirange testmeter with a.c. ranges, can be employed adequately for the measurement of voltage and current at power and audio frequencies. For measurements above about 10,000 Hz a valve-voltmeter or electronic testmeter is required, and such an instrument is considered in a later section.

Possibly one of the first measurements to be made with a new a.c. voltmeter will be the voltage existing at a mains power point. This may be done simply to check the instrument, or it may have another purpose, such as to ensure that power voltage is reaching the power input circuits of a receiver or amplifier. This kind of test would be made in the event of total failure of a piece of equipment, and when there is no visible evidence that the power supply is present

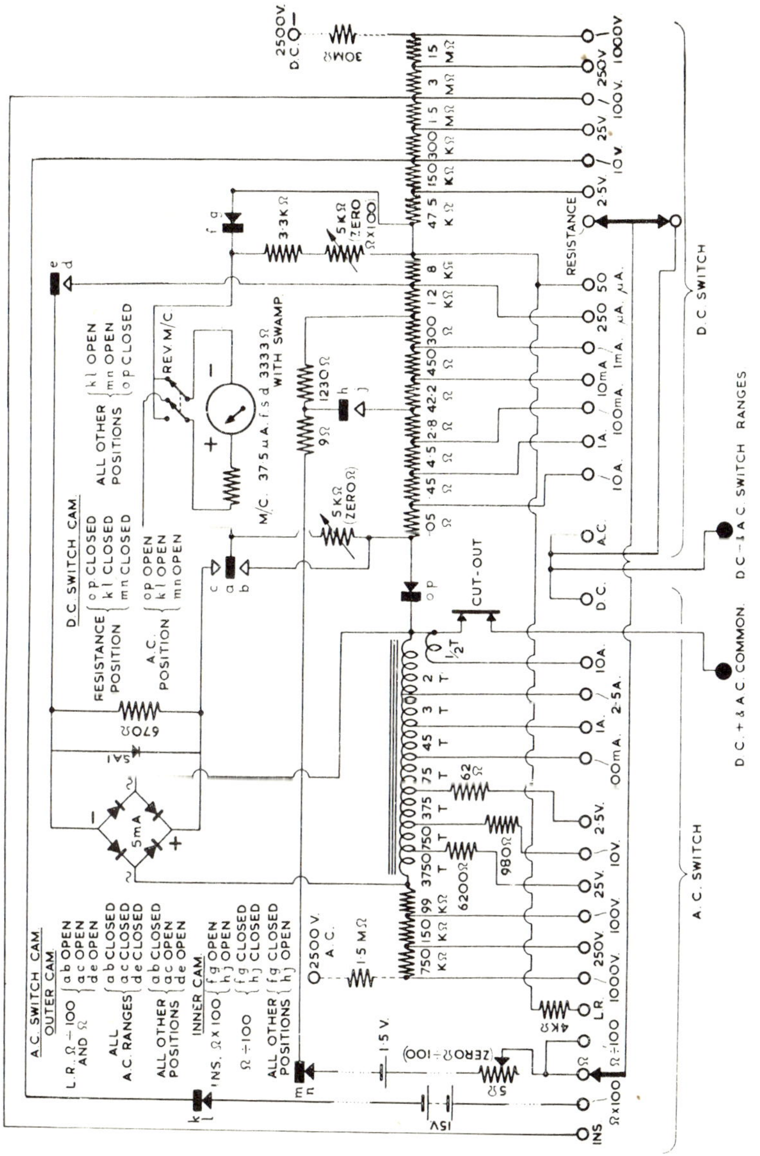

FIG. 3.5. *Complete circuit of the Model 8 AvoMeter.*

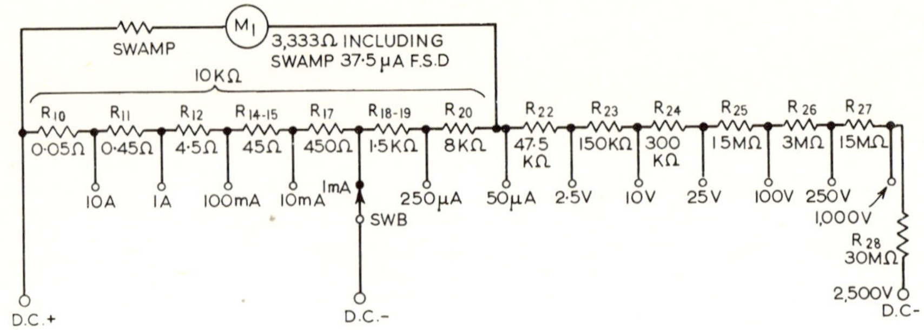

FIG. 3.6. *Simplified circuit of the d.c. ranges of Model 8 AvoMeter.*

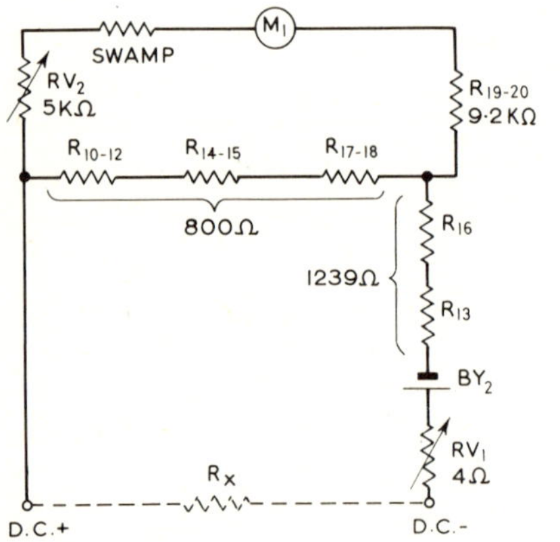

FIG. 3.7. *Simplified circuit of the ohms range of Model 8 AvoMeter.*

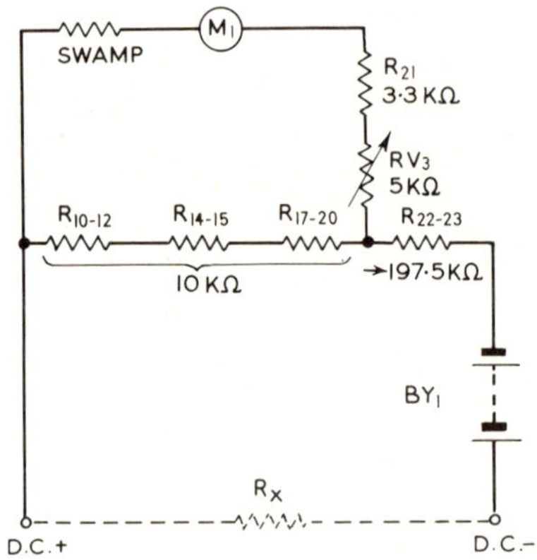

FIG. 3.8. *Simplified circuit of the × 100 range.*

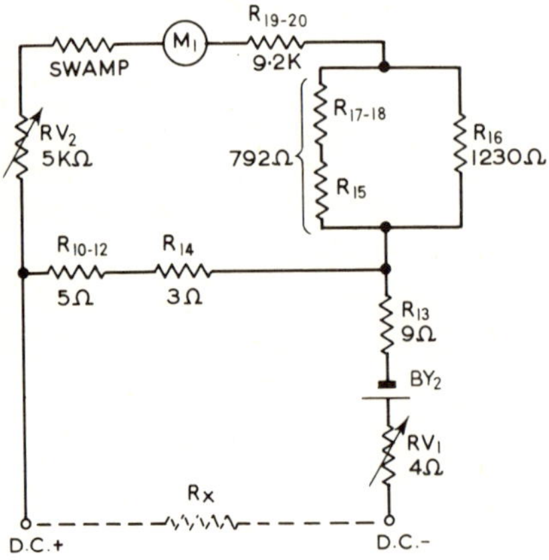

FIG. 3.9. *Simplified circuit of the ohms ÷ 100 range of the Model 8 AvoMeter.*

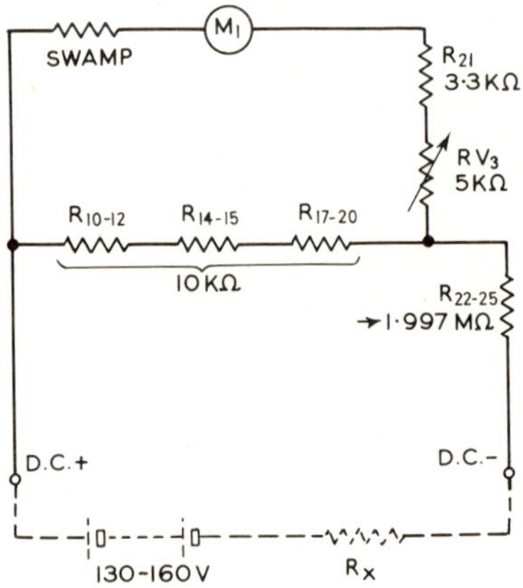

FIG. 3.10. *Simplified circuit of the insulation range.*

(e.g., if all the valve heaters and pilot lamp are not alight). The a.c. voltmeter would be adjusted to a suitable range (e.g., 250 V, 300 V or 150 V, according to the power-supply voltage and available ranges) and the power voltage would be traced from the supply socket to the set end of the power lead, past the fuses and the on/off switch and, if necessary, to the mains transformer, h.t. rectifier and valve heaters.

In the case of an a.c./d.c. type of chassis, an a.c. voltmeter adjusted to a range suitable for the power voltage can be used to analyse the power distribution, including checking the valve heaters, as shown in Fig. 3.11. If the equipment is dead, then one lead of the meter should be connected to chassis and the lead

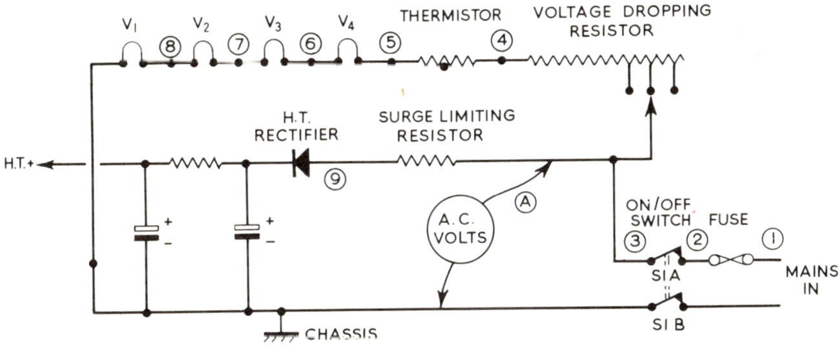

FIG. 3.11. *Checking the a.c. voltage in an a.c./d.c. type of receiver. The check points are numbered 1 to 9, and the voltmeter lead should be connected to each of these points in turn.*

marked *A* should be connected in turn to points (1) to (9). If there is no reading at point (1) and yet mains voltage is present at the set, section *S1B* of the on/off switch would be faulty. If voltage is present at point (1) but not at point (2), then the fuse would be open-circuited. All the items in the circuit can be checked in a like manner. For example, if voltage is present at point (6) but not at point (7), the V_3 heater is open-circuited. In this case, of course, the full mains voltage would be indicated at point (6), which would not be the case if the heater chain was intact, since there would be a voltage drop due to the current in the chain.

The a.c. voltage can also be checked right up to the input of the h.t. rectifier, point (9). Lack of voltage here would indicate open-circuiting of the surge-limiting resistor, which is a frequent cause of trouble in radio and television receivers. This trouble would not affect the heater chain, of course, and the heaters would be alight although the set would be dead. Beyond the rectifier the voltage is d.c., and tracing would have to be resumed with the meter adjusted to a suitable d.c. range.

The a.c. meter can also be used to measure the heater-chain current by breaking the circuit (say, immediately before valve V_1) and introducing the meter to complete the circuit. This check is often necessary if it is required accurately to adjust the tapping on the voltage-adjusting resistor. The heater current can be obtained from valve data sheets, and is the same for all valves in a series circuit. If the current is above that specified, then the valves are being over-run, or if below, the valves are being under-run; a corresponding adjustment to the dropping resistor will correct either of these defects.

A partial short-circuit in the heater of a picture tube in a television set can be determined by first measuring the heater-chain current to ensure that it is correct and then measuring the voltage across the tube heater. If the current is correct and the voltage across the heater low, then the tube heater has a partial short-circuit.

MEASURING POWER CONSUMPTION

The total power taken by any a.c. appliance or delivered to a load can be measured by first checking the a.c. voltage across it. This should be noted, and then a check of the total current should be made. Power consumption in watts is given by multiplying the voltage by the current. For example, if the voltage is 200 and the current 0·3 ampere, the consumption is 60 watts. The same applies to the measurement of d.c. power, but it should be noted that with a.c. the accuracy of power measurement made in this way is affected by the waveform deviating from sinusoidal and by the voltage and current being out of phase due to a reactive circuit element.

CHECKING AUDIO POWER WITH AN A.C. VOLTMETER

The audio output power of an amplifier or receiver can easily be measured by removing the loudspeaker and in its place connecting a resistor having an equivalent value. For instance, if the speaker impedance is 15 ohms, then a 15-ohm resistor should be used, of a wattage rating suitable to handle the expected amplifier output. An a.c. voltmeter should be connected across the resistor and the amplifier run to full output, preferably on a constant tone, such as that obtained from an audio generator or oscillator. The a.c. voltage reading should be noted, and the power output in watts is equal to the square of the voltage divided by the resistance. As an example: supposing the a.c. voltage to be 5; the square of this is 25, and 25 divided by the resistance, say, 5 ohms, is 5. So the audio output of the amplifier is 5 watts.

By feeding an amplifier with an audio signal of constant amplitude over a range of, say, 25 Hz to 10,000 Hz at suitable intervals and measuring the audio output at each setting, a curve can be plotted representing the frequency-response characteristics of the amplifier. Care should be taken not to set the audio input at such a high level that it is likely to overload the amplifier at certain frequencies. The curve is produced by plotting audio output against frequency, and several curves can be so constructed against different settings of the tone controls. See also Chapter 11.

MEASUREMENT OF CAPACITANCE

The reactance of a capacitor, usually indicated by X_c, is equal to $10^6/2\pi fC$, where f is the frequency in Hz and C the capacitance in microfarads. Reactance is measured in ohms, the same as with resistance, and in some respects exhibits similar properties in a.c. circuits as occur with resistance in d.c. circuits. It follows, therefore, that X_c is also equal to E/I, where E is the a.c. voltage applied across a capacitor and I is the current flowing through it.

Thus, by setting up a circuit as shown in Fig. 3.12 it is possible to determine the value of C with reasonable accuracy. To be strictly accurate the impedance (Z) of the circuit, which takes into account the resistance in the wires and losses in the capacitor, would have to be used; but since the resistance is small, in this

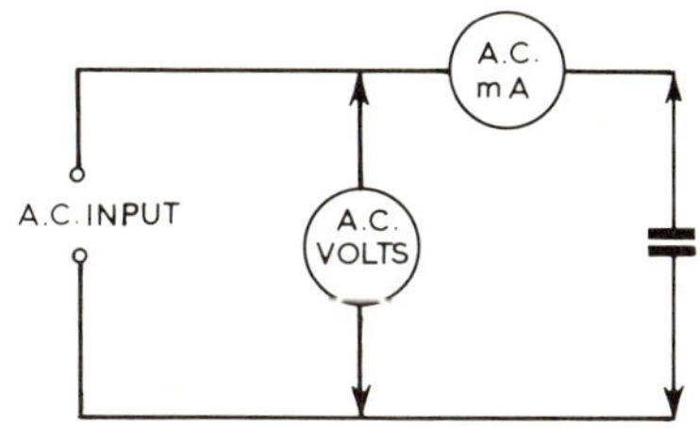

FIG. 3.12. *Circuit arrangement for the measurement of capacitance. It is advisable to disconnect the voltmeter when measuring the current.*

case the capacitive reactance may be taken as equal to the impedance. If readings of, say, 200 volts and 0·05 ampere (50 mA) are obtained, X_c equals 200/0·05, which is 4,000 ohms. Putting this in the other X_c equation we get $4{,}000 = 10^6/2\pi fC$, being the same as $C = 10^6/2\pi f \times 4{,}000$, which (at 50 Hz) works out to 0·795 microfarad. By combining the two X_c equations and converting the current to milliamperes, the following simplified expression results.

$$\text{Capacitance in microfarads} = F \times I/E$$

where F is a factor taking into account $10^6/2\pi f$ and the conversion to milliamperes, and is 3·18 for 50 Hz and 2·65 for 60 Hz; I is the capacitor current in *milliamperes*, and E the capacitor voltage.

When making this measurement it is best first to measure the voltage across the capacitor. Switch the circuit off, remove the voltmeter and introduce the current meter before switching on again. The domestic power supply can be used as the a.c. source, and since the frequency is usually controlled within fine limits, the accuracy of the measurement will be of a fairly high order, provided the resistance of the current meter is small compared with the reactance of the capacitor. Full isolation between the power supply and the circuit is desirable, and this can be achieved by the use of a double-wound transformer. A transformer having a tapped secondary is useful, since the voltage applied to the capacitor under test should always be a little below its maximum d.c. working voltage (0·7 times maximum). Also, with electrolytic capacitors, a very low voltage should be used (between 3 and 5 volts for h.t. smoothing units and reservoir capacitors). The reactance of low-voltage electrolytic capacitors cannot be measured easily by this method.

MEASUREMENT OF INDUCTANCE

In a similar way it is possible to measure the high-inductance values of iron-cored smoothing chokes, audio and power transformer windings, loudspeaker fields and so on. The inductor is connected in place of the capacitor in Fig. 3.12. The reactance of an inductor, indicated X_L, is equal to $2\pi fL$, where L is the inductance in henrys. X_L is also equal to E/I, where E is the applied voltage and I is the current. Thus, an equation similar to that for capacitance can be derived:

$$\text{Inductance in henrys} = F \times E/I$$

where F is 3·18 for 50 Hz and 2·65 for 60 Hz; I is the inductor current in *milliamperes*; and E the inductor voltage.

The accuracy in inductance measurements of this nature is slightly below that for capacitance measurements because inductors usually possess a relatively high contribution of pure resistance and a certain amount of winding capacitance. The current is thus limited by the *impedance* presented by the three elements and not by pure inductive reactance. The formula for impedance Z is $\sqrt{R^2+X^2}$, where R is the resistance and X is $X_L \pm X_c$. However, unless the resistance element is more than about 50 times the inductance in henrys, the

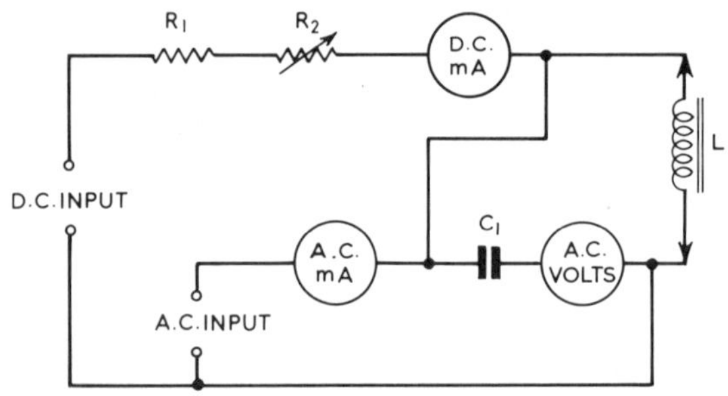

FIG. 3.13. *Circuit arrangement for the measurement of inductance with d.c. flowing in the test inductor L. Resistor R_1 acts as a d.c. limiter, and capacitor C_1 isolates the a.c. voltmeter.*

error by using pure inductive reactance will not be greater than about 1 per cent. At power frequencies, the winding capacitance will have negligible effect on the answer.

It should be remembered that the inductance of iron-cored chokes and transformer windings decreases when direct current as well as alternating current flows in the windings. This is due to the fact that the iron cores become partially saturated as a result of the magnetism produced by the direct current. This means that if, say, a smoothing choke is checked for inductance as described above, the figure obtained will not be the same as the inductance existing when the component is in actual service, since then it will have d.c. in the winding, as well as a.c.

A method of checking the inductance with d.c. flowing is shown in Fig. 3.13, in which R_1 is a d.c. limiter, R_2 is used to adjust the d.c. in the winding, and C_1 blocks direct current, so isolating the a.c. voltmeter. The capacitance of C_1 should be such that its reactance is small in comparison with the resistance of the a.c. voltmeter.

CHAPTER 4

Electronic Meters

ALTHOUGH modern multirange meters have high sensitivities, the power required to deflect the pointer over the scale may, nevertheless, often be in excess of the 'power producing' capabilities of the circuit under test. An indication is produced but the reading may be significantly below the actual voltage present at the test point with the instrument removed, as has been explained earlier.

Moreover, a.c. is more important now than in the past, from the servicing viewpoint. In addition to audio frequency, the service technician and the experimenter are often called upon to deal with video frequency, supersonic bias frequencies used in tape recorders, radio frequency and the very high frequency of the television channels. Multirange meters measure a.c., but usually meters with high-sensitivity on d.c. have a lower sensitivity on a.c.

In addition, the frequency range of the basic instrument is considerably restricted, rising to little more than 10 kHz at the best. It is evident, therefore, that something more elaborate than the multimeter is required for an ever-growing proportion of tests on radio, television and audio equipment, and the electronic meter is the next step up the ladder—at the top of which is the ideal but impossible meter with infinite impedance on both a.c. and d.c., a frequency response going up to 300 MHz and a reasonable indication at about 1 μV!

Electronic meters, therefore, are those in which active devices such as valves or transistors are employed to 'buffer', so to speak, the circuit under test from the readout, which can be an analogue meter-movement or a digital counter. There is a wide range of electronic type instruments, but that most extensively used is the valve or transistor multimeter which commonly provides facilities similar to those of the non-electronic multimeters, but often with range extensions and, of course, the attribute of significantly greater sensitivity, especially on the low voltage ranges. Although transistors are ousting valves, let us first take a look at the valve voltmeter, for there are many of these still in service.

Most valve-voltmeters have an input impedance of at least 10 or 11 megohms and a small amount of shunt capacitance, and some go as high as 25 megohms. A valve circuit is used as a buffer between the test circuit and the indicating instrument, and this supplies the power required to deflect the pointer. The test voltage or signal is thus 'looking into' the grid of a valve, which has very small effect on the circuit loading.

The conveyance of the test signal to the valve grid may upset matters, however, since the connecting leads can add inductance and capacitance to the test circuit. This can be remedied by extending the rectifier as close as possible to the point of circuit connexion and then running the rectified signals to the instrument through the normal leads, at which point capacitances are unimportant. Low-loss r.f. probes are available for this application, and measurement of a.c. voltages in excess of 200 MHz is possible.

The basic circuit of a valve-voltmeter is shown in Fig. 4.1. Two triode valves are arranged in the form of a bridge across the h.t. supply with the indicating

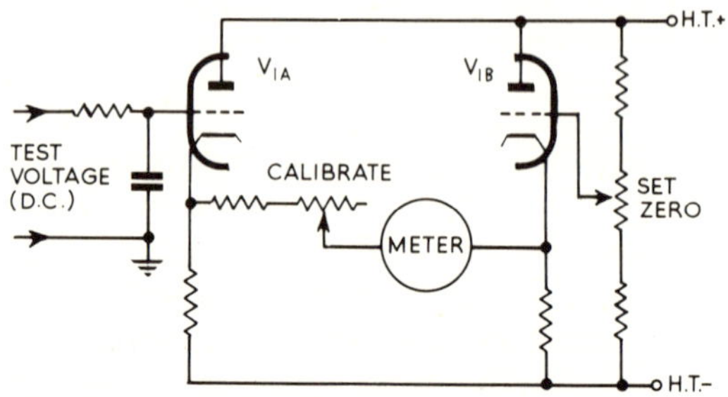

FIG. 4.1. *Basic circuit of a valve-voltmeter, showing how the triodes are arranged to form two arms of the bridge.*

meter connected between the cathodes. When each valve is passing exactly the same current, the voltage from cathode to cathode is zero and the indicating meter's pointer will indicate zero. Zero setting is obtained by adjustment of the bias applied to V_1B; a smaller or greater bias, relative to V_1A, will disturb the current balance in the valves and cause the pointer to deflect above or below the zero mark.

However, when the circuit is balanced and a test voltage is applied to V_1A grid, the balance will again be disturbed and the indicating meter will deflect. The extent of the deflection will depend on the magnitude of the test voltage, and the indicating meter is calibrated relative to the applied test voltage.

A.C. MEASUREMENTS

In order to measure a.c. voltages with this kind of instrument it is necessary to rectify the signal, and then apply it to the grid of V_1A in the form of d.c. A diode probe is usually available for this operation. The diode is housed in a small metal cylinder, one end of which is connected to the main instrument through a multi-core cable, and the other end is terminated by an insulated test probe. The diode and the probe are thus integral, and the diode can be connected very close to the signal source.

The circuit of a simple type of diode probe is given in Fig. 4.2. When an a.c. voltage is applied to the probe, the rectification action of the diode causes C_1 to charge so that the plate connected to the diode anode is negative and the plate connected to the probe is positive. Although this charge slowly leaks away through R_1, it is continually being renewed by the positive peaks of the applied signal. Therefore, a steady potential develops across C_1 and R_1 which is almost equal to the peak value of the signal or voltage waveform.

This potential is filtered and smoothed by R_2 and C_2 and is then applied to the grid of the valve-voltmeter valve in the same way as an ordinary d.c. voltage.

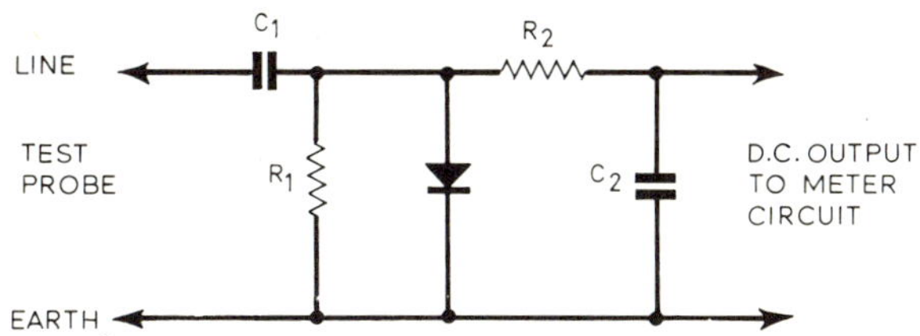

FIG. 4.2. *Circuit of diode probe for measuring a.c. with a valve-voltmeter of the type shown in Fig. 4.1.*

However, as the valve bridge responds to the peak voltage value of the waveform under test, the meter movement will deflect in accordance with this value, but since this is rarely required in practice, most movements are scaled in r.m.s. values, proportional to the peak input voltage. This means that the r.m.s. calibration on the meter will be correct only when sinusoidal voltages are being measured.

The basic circuit in Fig. 4.1 has been adopted in a valve multimeter by Avo in the 'Electronic Testmeter' whose circuit is given in Fig. 4.3 and the list of component values in Table 4.1. A multimeter of this kind is used in exactly the same way as the non-electronic multimeter, but a mains power supply source is required to energise the valves. In general, however, it will provide far more accurate indications in high resistance circuits than the non-electronic multimeter, while also facilitating the measurement of a.c. voltages at high frequencies, via the diode probe. This particular instrument incorporates a meter reversing switch, useful for maintaining a forward meter reading when performing a succession of d.c. measurements, the polarity of which is not consistent with respect to the 'common' test point. An inside view of the instrument is given in Fig. 1.3.

Contact Potential in Diode Probe

Owing to electrochemical activity between the electrodes of a diode valve, a small potential develops across the load even when there are no external voltages applied. This is called *contact potential* and, unless neutralised, would cause a small permanent deflection on the low a.c. ranges of a valve-voltmeter in which a diode *valve* is used in the a.c. test probe. It would also mean that if the instrument is set to zero on the d.c. ranges, the zero setting would not hold good on the a.c. ranges.

This effect can be eliminated by the use of a double-diode probe valve and arranging the contact potentials of the two sections to cancel out. Moreover, if the 'leak' resistor is returned to a slightly positive point in the circuit, then any

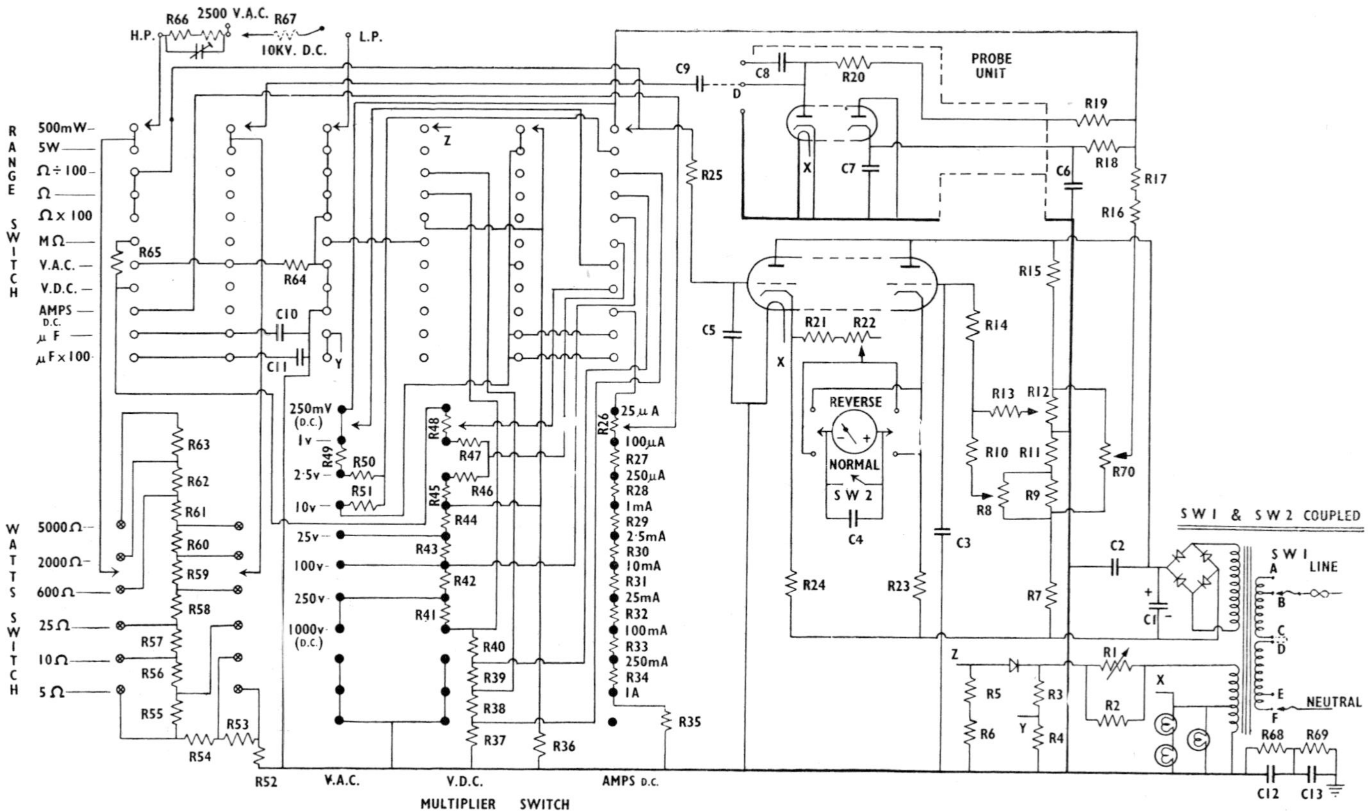

FIG. 4.3. *Circuit of early Avo Electronic Testmeter; component values are given in Table 4.1 opposite. The latest model (see Fig. 11.3) is solidstate.*

residual contact potential is cancelled out. Such an arrangement is adopted in the circuit of Fig. 4.3. The variable resistor, R_{70}, allows a small adjustment of the positive potential applied to the diodes, and thus serves as an 'a.c. zero' pre-set control. The effect, of course, is not present when the diode in the a.c. or r.f. probe is a semiconductor type.

Table 4.1

RESISTOR AND CAPACITOR VALUES

(see Fig. 4.3)

R	*Value*	*R*	*Value*	*R*	*Value*	*R*	*Value*	*R*	*Value*
1	100	18	7·5M	35	0·25	52	3·16	69	5M
2	36	19	Supp.	36	330K	53	1·31	70	$\frac{1}{2}$ to 1M
3	75	20	5M	37	16·6	54	0·53	*C*	*Value*
4	Supp.	21	1700	38	150	55	2·07		
5	9	22	1K	39	1·5K	56	2·93	1	16μF
6	9	23	500K	40	15K	57	15	2	0·1μF
7	80K	24	500K	41	50K	58	9·6	3	0·05μF
8	100	25	10K	42	100K	59	28·6	4	0·1μF
9	100	26	7·5K	43	500K	60	36·8	5	0·05μF
10	100K	27	1·5K	44	1M	61	500	6	0·05μF
11	450	28	750	45	825K	62	1·4K	7	470pF
12	500	29	150	46	550K	63	3K	8	470pF
13	50K	30	75	47	1·1M	64	825K	9	0·05μF
14	10K	31	15	48	8·25M	65	5M	10	0·0045μF
15	20K	32	7·5	49	1·93M	66	5·2M	11	0·5μF
16	15M	33	1·5	50	250K	67	95M	12	0·5μF
17	Supp.	34	0·75	51	615K	68	5M	13	0·5μF

In order to compensate for the non-linearity of a diode rectifier, the 'a.c. zero' pre-set is sometimes set so as to give a very small reading on the low a.c. voltage range. As this overcomes the diode non-linearity, all further a.c. measurements are read on a linear scale. Since this shift is very small it is barely noticeable on the high-voltage ranges.

Frequency Response Correction

The time-constant (R and C elements) of the diode probe has an influence upon the accuracy of measurement over a very wide frequency range. The time-constant is usually adjusted so that the instrument provides an accurate indication from power frequencies, through audio and up to low r.f. For greater accuracy at high r.f. a facility is usually available either allowing change of the capacitor or a slightly different operation of the probe.

Asymmetrical Waveform

When a simple diode probe is used for the measurement of a symmetrical waveform, such as a sine wave, the peak d.c. at the output of the probe will be the same whichever way round the probe is connected to the signal source. However, if the waveform is asymmetrical (for example, if one peak of the waveform is smaller or larger than the opposite peak), then the diode output voltage will be greater when it is passing, say, the large-amplitude positive peaks than if it were connected to the source the other way round, when it would then be passing the small-amplitude negative peaks. This is an important point to have in mind when using a diode probe on a waveform which is likely to deviate from a pure sine wave.

Input Resistance and Impedance

The input resistance of a d.c. valve-voltmeter is equal to the total resistance forming the multiplier or voltage divider arrangement. The test voltage is usually applied directly across this resistive system, as shown in Fig. 4.4. This means that the input resistance remains the same over all the ranges, which is

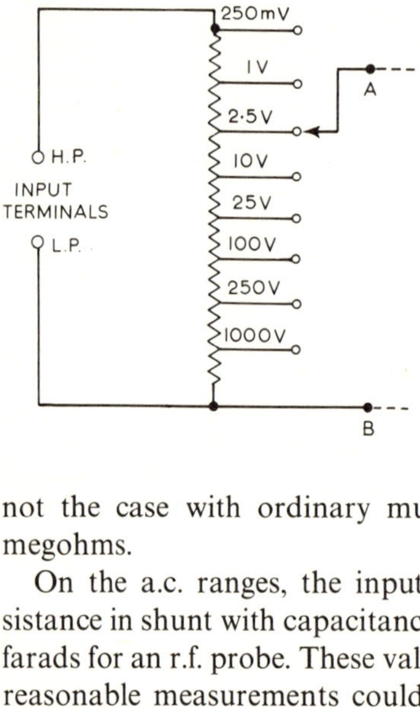

FIG. 4.4. *The 'd.c. volts' measuring circuit of the Avo Electronic Testmeter (see Fig. 4.3).*

not the case with ordinary multimeters. A common input resistance is 11 megohms.

On the a.c. ranges, the input characteristic is sometimes expressed as resistance in shunt with capacitance. Typical values are 1·75 megohms and 7 picofarads for an r.f. probe. These values would fix the maximum frequency at which reasonable measurements could be made at approximately 200 MHz.

Testmeters using Transistors

We can glean a very good impression of a simple transistor voltmeter from the circuit in Fig. 4.5. This, using a field-effect transistor (f.e.t.) is from a Mullard Educational Project in Electronics. It will be seen that the transistor is arranged as a source-follower, meaning that the output for operating the meter movement emanates from the source electrode of the f.e.t. The input, of course, is applied to the gate electrode, the resistance of which is incredibly high. Indeed,

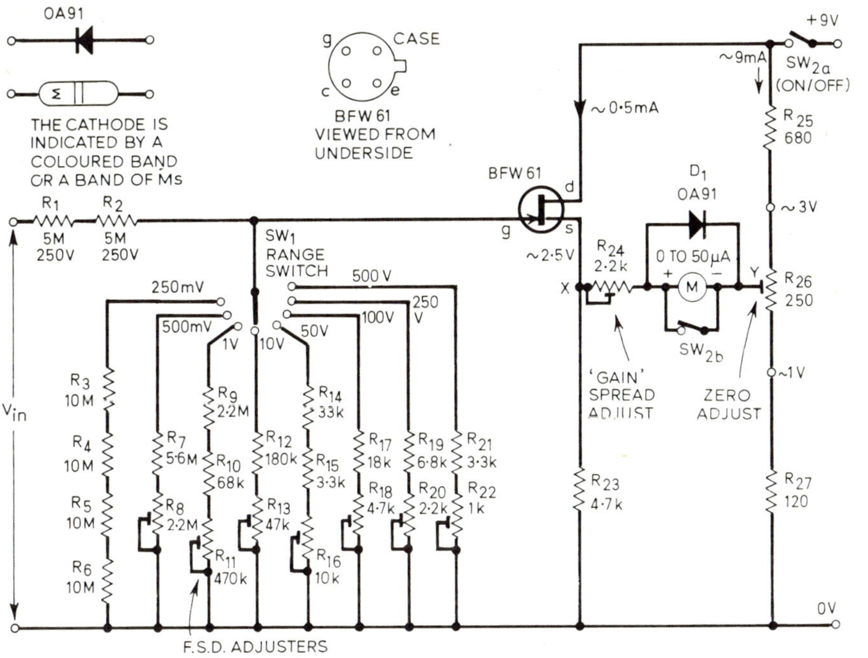

FIG. 4.5. *Circuit diagram of a simple f.e.t. voltmeter by Mullard. This is fully described in the text.*

the effective resistance between the gate electrode and the 0 V line is so high that its shunting effect on the switched-in resistor network is negligible, even on the lowest range (250 mV). This very high input resistance is enhanced by the large negative feedback which stems from the f.e.t. configuration.

The drain of the f.e.t. is connected direct to the 9 V rail, and the drain/source idling current is in the order of 500 μA in the absence of input voltage. This yields a potential of about 2·5 V at the source (point X on the circuit), but f.e.t. spreads could modify this within a range of about 1 to 3 V. The plan is to get the meter to read zero in the absence of input voltage and this is achieved by the potentiometer R_{26}, padded one side by R_{25} and the other side by R_{27} across the supply circuit. R_{26} is thus adjusted until the potential at its slider equals that at the f.e.t. source. Under this condition the bridge arrangement is balanced and no current flows through the meter. Variable resistor R_{24} accommodates f.e.t. spreads and facilitates calibration.

The meter movement is protected against overloads by diode D_1, while switch SW_2b shorts the movement out when the on/off switch SW_2a is in the 'off' position, thereby heavily damping the movement and inhibiting pointer oscillation in transit.

The input attenuator operated by switch SW_1 consists of a potential divider to the gate electrode, where $R_1 + R_2$ is in the top leg and the series chain of resistors selected by SW_1 in the bottom leg. The circuit is designed so that the potential at the gate relative to the 0 V line is 200 mV for all f.s.d. inputs.

It will be seen that all but the 250 mV chain of resistors at the input switch include a 'trimming' resistor preset. After adjusting R_{26} for meter zero, the preset appropriate to the switched range is adjusted for meter f.s.d. when an accurate calibrating voltage is applied at the input. Since a f.s.d. preset is not incorporated for the 250 mV range, the f.s.d. adjustment on this range is handled by the f.e.t. 'spread' preset R_{24}. In fact this range is adjusted first to establish the correct operating conditions.

Method of calibration as suggested by Mullard is as follows.

1. Switch to the 250 mV range and adjust R_{24} to minimum (zero ohms). Remove calibration source, short input terminals together and then adjust R_{26} for zero meter deflection. Remove input short.
2. Apply 250 mV input at the correct polarity (see Fig. 4.5) and adjust R_{24} for f.s.d.
3. Repeat steps 1 and 2 until the readout is correct at both ends of scale. *After this R_{24} should not be re-adjusted.*
4. Switch to the 500 mV range and check that zero deflection is obtained with the input shorted (source disconnected of course!). If this condition is not achieved it will be necessary to repeat steps 1 to 3 again. Remove input short, apply 500 mV and adjust R_8 for f.s.d.
5. Switch in turn to the other ranges with suitable f.s.d. calibration voltage and adjust the associated presets R_{11}, R_{13}, R_{16}, R_{18}, R_{20} and R_{22} to secure f.s.d. on each range.

Although negative feedback in the f.e.t. circuit ensures good linearity within the range of zero to 200 mV, since this limits the operation to a diminutive section of the I_d/V_{gs} characteristic, the accuracy is ultimately dependent on the accuracy of the resistors and on the setting of the presets. Mullard intimate that this particular design is not meant specifically to reflect an 'accurate piece of test gear'. Nevertheless, it shows admirably how a transistor can be exploited in electronic voltmeter design. The f.e.t. is particularly suitable for this application owing to its high gate input resistance, especially when arranged in source-follower mode.

APPLICATION OF THE ELECTRONIC METER

When valves are used in the instrument, sufficient time should be allowed, before putting the instrument into use, for the valve circuits to become perfectly stable, which is shown when there is no longer any drift of the pointer away from the zero mark. If there occurs a steady movement of the pointer away from zero, even after the instrument has had time to warm up, then there is either a faulty valve or an open-circuited multiplier.

D.C. testing is performed in the usual manner, after first selecting the required range. The very high input resistance enables tests to be made with great accuracy on high-resistance circuits, such as on a.g.c. lines in radio and television sets, in f.m. discriminator circuits and ratio detectors, on high-resistance feed circuits to the anodes and screens of valves in amplifier and time-base circuits, and so on.

There are usually one or two precautions that should be taken when using this type of instrument. For instance, there should be no high potential difference between the instrument negative lead and the chassis of the instrument, which should be at true earth potential. A very high potential difference between these two points would result in arcing or insulation breakdown within the instrument.

If r.f. is present on the d.c. test voltage, an r.f. stopper resistor should be connected between the voltage source and the instrument test lead. This will affect the voltage reading, depending on the value of the resistor used, but can be taken into account in arriving at the true d.c. voltage. With some instruments an external multiplier resistance is available, which serves to extend the range on d.c. volts, and this can be employed also as an r.f. stopper.

High-frequency Measurements

The a.c. probe can be used in conjunction with the measurement of the voltage of the r.f. signals in a receiver up to the point of the detector. It is unlikely, however, that the instrument will be sensitive enough to give a usable indication of the very small voltage before the frequency-changer stage. Nevertheless, with a strong aerial signal, the i.f. at the collector of the frequency changer should be measurable in terms of millivolts. The signal can then be followed through, checking at the collector of the following i.f. stages and finally at the detector diode. This represents a form of signal tracing, which is more desirable from a constant-input signal such as can be obtained from a signal generator, and which is dealt with in the next chapter.

The probe should always be connected to the r.f. circuits through the shortest possible length of wire. Where the probe is connected across a tuned circuit, the trimmer or tuning capacitor should be adjusted, where possible, to compensate for the extra shunt capacitance of the probe input circuit. This is necessary when making absolute measurements of stage gain.

The valve-voltmeter is extremely useful for measuring the amplitude of the signal at the local oscillator of radio and television receivers, and to make sure that the oscillator is working in the event of a receiver fault. The 'live' side of the probe should be connected to the fixed plates of the oscillator tuning capacitor or, in the case of television receivers, to the grid of the oscillator valve, while the 'earth' side of the probe should be connected to receiver chassis. An indication between 2 and 25 volts is usually obtained, and with valve radios will alter as the tuning gang is swung from minimum to maximum capacitance. With television sets, the oscillator voltage may well alter over the different channels. On Band III, or higher frequency bands, the capacitive loading of the probe may disturb the circuit so much as to stop oscillations.

In the case of excessive loading, as described above, the operation of the oscillator can be assessed by using a d.c. valve-voltmeter connected between oscillator-valve grid or transistor base and chassis.

Checking R.F. and A.F. Decoupling

Decoupling capacitors in r.f. and audio amplifiers serve to by-pass any r.f. or a.f. voltage at their point of connexion direct to earth or chassis. It follows, therefore, that such capacitors should have a very low impedance at their operating frequency, and thus a very small or negligible r.f. or a.f. voltage developed across them. However, if the capacitor is low in value, or open-circuited, then the radio-frequency or audio-frequency potential across it will be high, and this can be checked with an a.c. valve-voltmeter.

It should be noted, though, that apparently inefficient decoupling is sometimes purposely arranged to provide degeneration (i.e., negative feedback) at certain frequencies. Capacitors in the cathode circuit of some television video amplifiers are arranged in value with this in mind. An unusually high voltage developed across such capacitors would not necessarily indicate that the circuit design is wrong or that the component is faulty. Similar arrangements may be found in negative-feedback loops of audio amplifiers and radio receivers.

Audio Applications

An a.c. valve-voltmeter can easily be utilised as an output meter or to measure output power, as described in Chapters 3 and 11. It can also be used to trace the audio frequency signals through an amplifier, from the signal source to the loudspeaker.

The overall balance of a push-pull audio stage can quickly be checked by measuring the voltage applied between chassis and each output-valve grid. This is best done with the amplifier receiving a constant signal from an audio oscillator. The voltages should be alike if the phase-splitter stage is correctly balanced. Similarly, like voltages should be obtained between each output-valve anode and chassis if the output valves and circuits are balanced. The same general technique can be carried back to preceding driver stages if used in the push-pull mode.

When a diode probe is used for the measurement of relatively low-value a.f. or r.f. voltages, but superimposed upon these voltages is a high d.c. voltage—such as may be experienced at the anodes of high-power audio output stages—then as a safeguard against probe failure a protective capacitor of suitable value, and having the necessary d.c. voltage working range, should be connected as a d.c. stop between the 'live' terminal of the probe and the test circuit. The value should be chosen so that the capacitor does not offer undue reactance at the test frequency.

AUDIO MILLIVOLTMETER

This differs from the electronic voltmeter or multimeter in that it is designed essentially for the measurement of a wide range of a.c. voltages, including extremely small voltages, over a relatively extended frequency spectrum. Some such instruments have a response from 15 Hz to 10 MHz, while others may not go as high as this, depending on the design.

The instrument consists basically of a wideband amplifier and switched attenuators to give the various ranges. A two- or four-stage amplifier, either in conventional mode or in the form of cascode stages, with liberal negative feedback, feeds a bridge rectifier, whose d.c. output voltage operates a sensitive moving-coil meter calibrated in r.m.s. a.c.—sometimes with an extra scale in decibels (with 0 db relative to 1 mW into 600 ohms).

These instruments are endowed with a very high sensitivity, some reading down to 100 μV f.s.d. or less. Although designed primarily for audio-frequency applications, some models are suitable for measurements at low r.f. and at video frequencies. Those models with very low f.s.d. ranges are sometimes called microvoltmeters.

For the audio workshop and test laboratory an audio millivoltmeter is essential; it can quickly determine, for example, the output from pickup cartridges, microphones and tape heads. Ripple voltage, a.c. power voltage, noise voltage from amplifiers and the like can also be measured, and as most species incorporate a dB scale or scales it is a simple matter to check the frequency response of an amplifier or other item of audio equipment using dB ratios on the Y axis and frequency on the X axis, the latter to a logarithmic scale.

Some models are designed so that the internal amplifier can be exploited for external application, independently of the meter movement or together with it. This can be useful for increasing the Y sensitivity of an oscilloscope, for example. Moreover, the signal amplified by the instrument can be observed on the screen of an oscilloscope, and this is particularly useful when the millivoltmeter is reading distortion components from a distortion analyser as it allows the nature of the distortion to be examined (see Chapter 11).

As with most other instruments of this kind, the calibration is commonly in r.m.s. voltage based on a sinewave input.

The circuit of a recent transistor design is given in Fig. 4.6. This is another one of the J. E. Sugden instruments, of which there is a trio consisting of the Si 453 audio oscillator referred to in Chapter 1 (see Fig. 1.7), a distortion measuring unit, Model Si 452 (see Chapter 11) and the millivoltmeter (Fig. 4.6) which is Model Si 451. Noteworthy is the fact that all these instruments employ transistors and are battery powered; the latter, in conjunction with insulated feet, avoids many of the problems associated with hum-loops and r.f. couplings when tests are being made in sensitive and powerful amplifiers.

The major part of the circuit in Fig. 4.6 consists of signal amplification, using bootstrapping in the first X30 three-transistor amplifier and adequate feedback in both this and the X270 three-transistor amplifier for purposes of low dis-

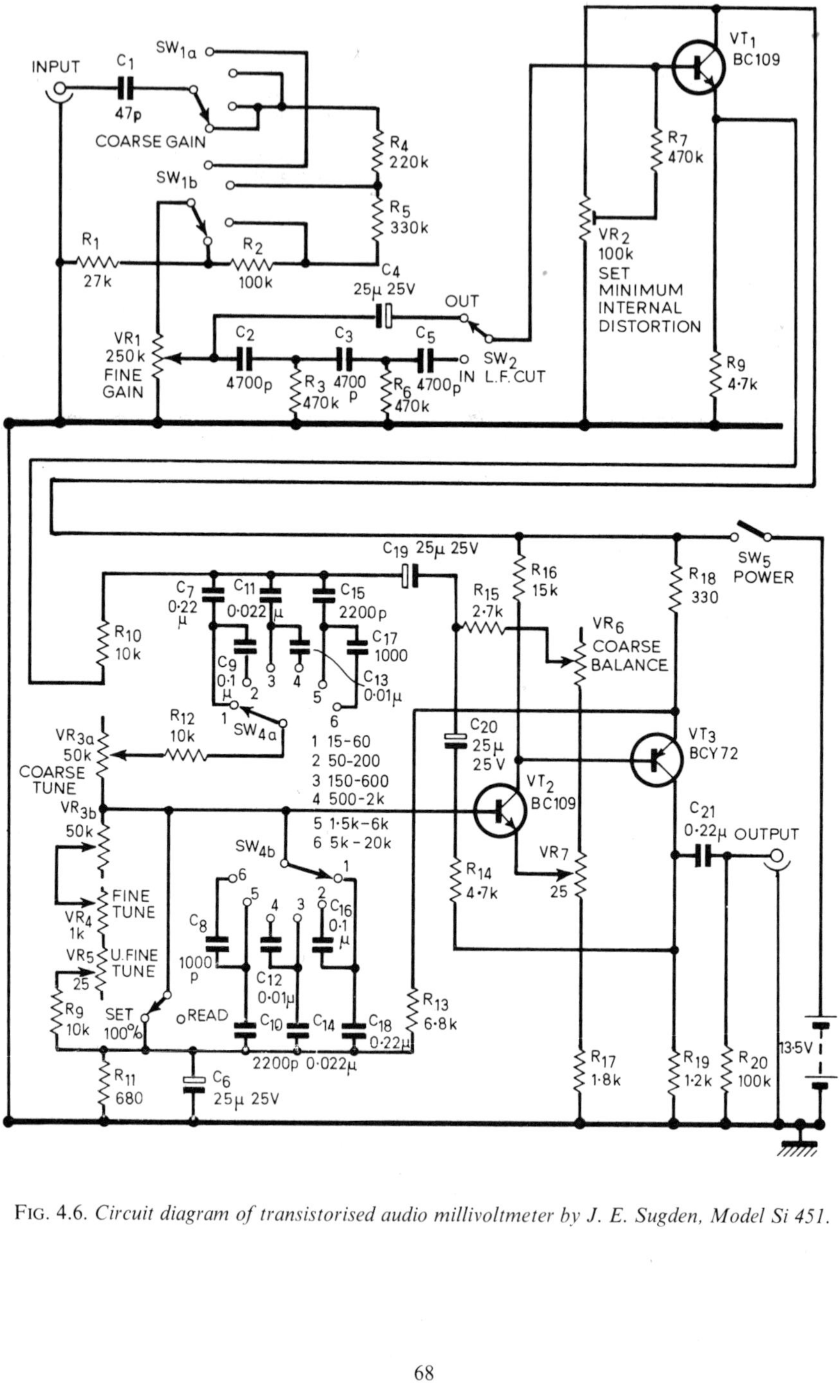

FIG. 4.6. *Circuit diagram of transistorised audio millivoltmeter by J. E. Sugden, Model Si 451.*

tortion and high stability. The input attenuator sets the range of the instrument, while the attenuator between the X30 and X270 amplifiers adjusts the readout to the appropriate scale on the meter. The meter is energised by way of the final transistor VT_7, the amplified audio signal, of course, being rectified prior to the movement by X2.

It will be seen that the amplified signal is available at the outlet connected from the emitter of VT_6, via R_{30}, C_{14} and C_{15}.

The potentiometer VR_1 between VT_3 and the scale attenuator is arranged so that at one extreme the readout is in r.m.s. voltage and at the other extreme in peak-to-peak voltage. Since the maximum sensitivity is 1 mV r.m.s. f.s.d., the peak-to-peak attribute means that with the control set for that condition of measurement the effective sensitivity is increased to about 350 μV r.m.s. f.s.d. Moreover, the control is convenient for establishing a suitable dB datum when making frequency response and similar measurements, for when such is the exercise the engineer or technician is primarily concerned with *relative* deflection based on the dB scale and not with absolute values of voltage. The 'open' portion of the dB scale, corresponding to 20 dB, is established close to 80 per cent of the f.s.d., thereby making small dB changes easily visible when the

FIG. 4.7. *Avo Digital Avometer Type DA.122. This is designed for mains or battery operation, the latter from an internal rechargeable battery allowing measurement without earth-loop errors. Facilities can also be obtained for operating from an external 12 V or 24 V supply. Design is solid-state and measurements are a.c. and d.c. voltage, current and resistance. D.C. voltage ranges have an overall capability of 100 μV to 1,500 V with a 50 per cent over-range facility and an accuracy of 0·1 per cent of reading ±0·1 per cent of full range on the 1 V and 10 V ranges. Measured value is indicated on a 3½ digit display with auto decimal point and over-range indication.*

datum is referred to 20 dB on the scale (0 dB equals 1 mW into 600 ohms on the 10 V r.m.s. range).

As just noted, maximum r.m.s. sensitivity is 1 mV; minimum is 20 V r.m.s. Frequency response is to 0·5 dB over the range of 20 Hz to 20 kHz. The range is in terms of 1, 10, 100 and 1,000 and the scale in terms of 1, 2, 5, 10 and 20. Position BC on SW_2B of the scale switch (at the meter) is a battery check position, this merely allowing the meter to monitor the battery voltage via R_{36}.

Input impedance is 1·1 M and the output signal available is 3 V r.m.s. at f.s.d. from a low impedance source (e.g., emitter of VT_6). In common with the other instruments of the trio, power is supplied by three internal batteries each of 4·5 V, such as the Ever Ready AD28 or Exide H1176, and stabilised by VT_8.

CHAPTER 5

Signal Generators

SIGNAL GENERATORS are instruments for producing a controlled source of signal for test purposes over a specific frequency spectrum. There are two basic types: the a.f. generator or oscillator, which produces signals within the audio range and sometimes above; and the r.f. generator, which produces signals within the radio-frequency spectrum. The basic specifications of such instruments are given in Chapter 1.

R.F. SIGNAL GENERATOR

This instrument produces a continuous wave signal of known and variable frequency and amplitude which, for the purpose of general servicing, can be modulated to a known depth to simulate the signals that the equipment under service would normally receive, either from an aerial or other signal source.

Modulation

R.F. signal generators are designed basically to yield either an amplitude modulated carrier wave with the smallest possible residual frequency modulation, or a frequency-modulated carrier wave with the smallest possible residual amplitude modulation. For ordinary servicing they are sometimes designed to provide either a.m. or f.m. as required, by the turn of a switch. The modulation signal is usually an a.f. sine wave of 400 or 1,000 Hz, but in addition a facility may be available for square-wave modulation. The modulation signal can usually be made available, for audio tests, at two terminals on the front panel. It is also sometimes possible to switch off the internally generated modulation and to modulate the carrier by a signal from an external source, such as an audio generator or square-wave generator.

Some instruments include a modulation-depth indicating meter, and a control allowing the depth of either the internal or external modulation to be adjusted as required. Where this facility is not provided, however, the modulation depth is generally set internally to about 30 per cent, which is a value suitable for many tests. All instruments contain a modulation on/off switch, allowing tests to be made with the carrier alone. Even though modulation depth control might not be provided, there is usually an a.f. level control, which permits adjustment to the level of the a.f. at the a.f. output terminals. This control is commonly calibrated roughly in terms of audio-frequency r.m.s. voltage.

Video Modulation

A somewhat specialised type of signal generator, usually called a 'pattern generator', uses modulation of a waveform which is representative of the picture and sync pulses of a television signal. The pattern of this modulation signal is arranged in such a way that when the modulated carrier is applied to a television receiver there appears on the screen a 'test signal' which ranges from peak white to black. This is made up of a series of vertical bars and horizontal lines, thus allowing a television receiver to be tested or set-up in the absence of a real transmission.

Modulation for Colour Sets

As detailed later, a new type of pattern generator has appeared since the first edition of this book. This is called a Colour Pattern Generator since the video modulation that it produces is tailored to facilitate specific tests and adjustments in colour television sets. Some models provide a video signal that is switchable to yield a crosshatch or dot pattern on the screen and a white raster for purity adjustments, sometimes with the extra facility of contrast bars for grey-scale tracking, while more expensive instruments provide a fully colour-encoded signal in addition for producing colour displays (sometimes the standard colour bars or a rainbow pattern) so that tests and adjustments appropriate to the receiver decoder sections and colour circuits can be performed with the least effort and maximum accuracy. The simple type of crosshatch/dot generator is mainly designed for static and dynamic convergence adjustments, the dot pattern for the former and the crosshatch for the latter.

Carrier Control

The carrier wave, either modulated or unmodulated, as required, is available, usually at an impedance of 75 ohms, at one or two sockets on the front panel of the instrument. On one socket, the signal is passed through a variable attenuator system so that its level can be adjusted to suit the test in hand. On the other socket, which is not fitted to all instruments, the signal is passed direct from a high-level point in the oscillator. This is sometimes called the 'force output', and such a high-level signal may be useful in forcing a signal through a completely misaligned receiver to allow the initial adjustment of one or two trimmers, after which the test signal can be supplied via the attenuator in the normal way, and the signal output reduced as the remaining trimmers are adjusted to bring the circuits into tune.

On some instruments, the r.f. signal applied to the input of the variable attenuator system is monitored on a small indicating meter. There is usually a red mark or other reference point to which the needle should be set by adjusting the 'r.f. output control' before commencing operations. This ensures that the correct r.f. is always applied to the input of the attenuator, for it sometimes happens that the r.f. alters in amplitude as its frequency is changed, or as the attenuator is loaded. Correction can thus be made to the r.f. output over the whole frequency range of the instrument.

Attenuator System

The attenuator system provides a continuously variable source of e.m.f. from about 1 microvolt to the maximum r.f. output of the instrument, which in most service-type instruments is 100 millivolts. The minimum r.f. output level is a function of the design of the attenuators and the screening of the instrument, for the residual radiation from the instrument itself may exceed 1 microvolt. With expensive laboratory-class instruments, detailed attention to the attenuators and screening allows for r.f. control down to about 0·1 microvolt.

There is usually a 'step attenuator' which enables the r.f. to be reduced in steps of 20 db (10 times) down to about −80 db, and a continuously variable attenuator which allows each step to be adjusted by a further 20 db. Thus, if the step attenuator is set at, say, −20 db and the variable attenuator also to −20 db, then the total attenuation between the output of the oscillator and the r.f. output socket is 40 db, or 100 times. If the maximum output is 100 millivolts, then 40 db attenuation will give a terminal voltage of 1 millivolt.

In addition to decibels, some attenuators are also calibrated in terms of actual output voltage, relative either to the full output of the instrument or to the setting of the r.f. control and indicator. In this case the signal level given on the step attenuator is multiplied with the signal level given on the continuously variable or 'fine' attenuator. Thus, if the step attenuator is set at, say, 1 millivolt and the fine attenuator to position 5, then the r.f. output is 5 millivolts. When so used, the step attenuator is often called a 'multiplier' and is calibrated in microvolts × 1, × 10 and × 100 and in millivolts × 1 and × 10 and the fine attenuator is calibrated from 1 to 10. It is important to remember that when dealing in signal volts multiplication is required, whereas the two values are added when they are in terms of decibels.

Frequency Range and Tuning

It is becoming customary for service-type r.f. signal generators to embrace the whole of the r.f. spectrum from around 100 kHz to 240 MHz in eight or so ranges. This makes the instrument suitable for service work on broadcast receivers, on f.m. receivers and on television receivers (Bands I, II and III), including the i.f. bands. On some instruments, such a great range is facilitated by using the second or third harmonics of the lower range signals to provide the signals of the higher ranges. Other manufacturers feel that the use of harmonics makes life difficult for the service technician, and an endeavour is therefore made to cover the whole spectrum on fundamentals. There is little doubt that fundamental frequencies are far easier to use than harmonics, and are less likely to cause errors due to spurious receiver responses which cannot quite be tied to the frequency at which the generator is adjusted.

Certain tests in dual-standard and single-standard (625 lines only) monochrome and colour receivers require a u.h.f. generator tuning over the channels in Bands IV and V, and this applies particularly to colour pattern generators. Early models were designed to tune over the v.h.f. channels in Bands I and III as well as over the u.h.f. channels in Bands IV and V, but since the advent of the single-standard (625-line-only) receiver the trend is for the tuning to cover the u.h.f. bands only.

As u.h.f. tuner alignment operations are rarely carried out by the service technician, the general plan being to replace the u.h.f. tuner and return the suspect or faulty one to the maker for repairs or realignment, a u.h.f. service generator is not truly an essential instrument since the v.h.f. generator can usually be brought into service for 625-line standard i.f. alignment. However, an f.m. generator tuning over 6 MHz for sound intercarrier channel alignment is essential nowadays, and it is desirable to have some means of checking the frequency since misalignment of the intercarrier sound channel or the f.m. ratio detector transformer is commonly responsible for sound distortion. Some generators have a crystal check facility for this (see the section which follows under 'Crystal Check'), but failing this the generator signal can be adjusted against the 6 MHz sound intercarrier signal present in a correctly aligned receiver by tuning to the 'dead beat' point (see Fig. 1.11).

There are various types of tuning dials, and in spite of the wide frequency range covered, tuning is not unduly critical as the mechanical spread of the dial is equal on some instruments to a total length of 50 inches. Resetting to a particular frequency is also facilitated on some instruments by an auxiliary incremental dial, independently calibrated and driven from the main dial. Such a feature also allows for easy interpolation for bandwidth measurements and so on.

Other instruments incorporate a bandspreading arrangement and an adjustable cursor, which enables the instrument to be set against standard frequencies; e.g., broadcast carrier signals or a crystal calibrator. On most models, the normal accuracy of calibration is of the order of ± 1 per cent.

A few service-type instruments may not cover such a wide frequency spectrum, in which case two or more instruments may be required in an average service department to provide the necessary frequency range.

Crystal Check

Service generators with an inbuilt crystal oscillator for checking the accuracy of calibration at certain points on the dial also have much to commend them.

The crystal controls the frequency of a separate oscillator which is rich in harmonics. If, for instance, a 1-MHz crystal is used, when the oscillator is switched on there will be produced not only the 1-MHz fundamental signal, but also other signals at intervals of 1 MHz throughout the normal tuning range of the instrument. These check-frequency points can be arranged to beat with the tuned frequency of the main oscillator, so that the actual calibration can be assessed within very fine limits every 1 MHz, or depending upon the crystal frequency and type of circuit employed.

Some instruments have facilities for detecting the two signals, and then amplifying the beat-note so that it can be clearly heard in a pair of headphones, for which a jack socket is provided on the front panel. Such a beat-note is illustrated diagrammatically in Fig. 1.11(*b*).

Dummy Aerial

When using a signal generator to simulate such a signal as would be applied to a receiver from an aerial, it is important that the loading effect of the generator is as near as possible to that of the aerial itself. This is particularly important when making sensitivity and other tests on broadcast receivers with the generator signal applied to the receiver's aerial socket.

The output impedance of most signal generators is in the region of 75 ohms, and a coaxial termination is invariably fitted to the front panel. Such an impedance matches exactly the aerial-circuit input impedance of television and recent v.h.f./f.m. receivers. Thus, the generator may, when required, be connected direct to the aerial input socket through a length of coaxial cable.

However, with medium-frequency broadcast receivers the aerial-circuit input impedance is relatively high, since this 'sees' a high impedance at the end of an ordinary aerial. An ordinary aerial, although of high impedance, possesses distributed resistance, inductance and capacitance, and these elements are also reflected to the aerial input socket under normal operating conditions. It follows that such elements should also be present even when the signal is obtained from a signal generator.

These are provided by a dummy aerial which is, in effect, a terminating unit supplied with the signal generator. This changes the low impedance of the generator signal source to a high impedance while also providing resistance, inductance and capacitance in correct proportion, or in the proportions to match an 'average' aerial. Thus, when the generator signal is applied to such a receiver through a dummy aerial the receiver cannot distinguish between it and that which would normally be applied from an aerial. This means, of course, that when the alignment is completed no correction is necessary to work the set from an ordinary aerial.

The circuit of a dummy aerial suitable for medium frequencies is given in Fig. 5.1*a*. To simulate the characteristics of a car-type aerial, correctly mounted on the body of the car, the circuit of 5.1*b* is recommended, while the circuit of 5.1*c* is suitable for general frequency coverage when used in conjunction with ordinary broadcast receivers.

The photographs in Fig. 5.1 show external and internal views of a dummy aerial at *d* and *e* respectively. B.S. 4054:1966 gives details of other types.

When a signal generator and dummy aerial are connected to a receiver with a chassis not isolated from the mains (e.g., most a.c./d.c. sets), the set should be plugged in with a polarity on the supply socket such as to produce

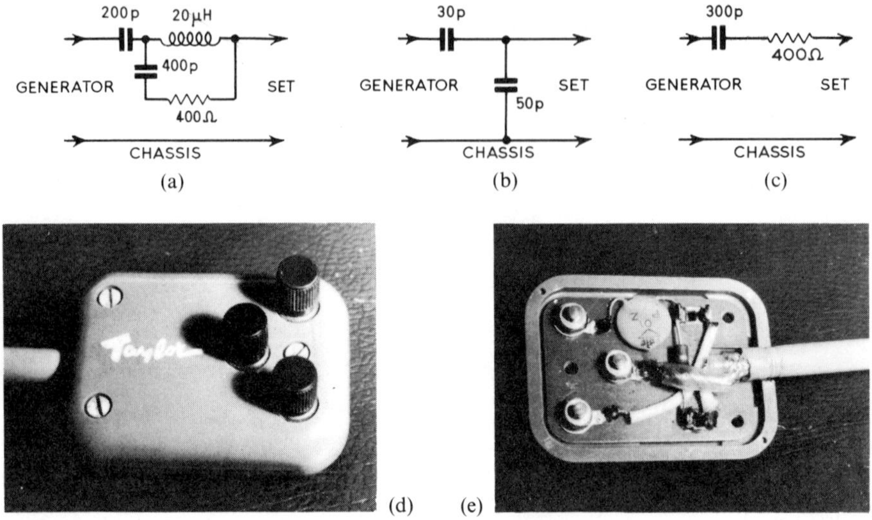

FIG. 5.1. *Dummy aerial configurations,* (a) *for medium frequencies,* (b) *for car radios and* (c) *for general purposes. External and internal views of a typical dummy aerial are shown at* (d) *and* (e).

'neutral' mains voltage on the chassis. Tests via the receiver's aerial socket may then be undertaken reasonably satisfactorily by interposing an ·0·05 microfarad paper capacitor in series with both the 'live' lead and the 'earthy' lead from the dummy aerial. (See also the section on 'Safety Precautions' in Chapter 1.)

A.M. SIGNAL GENERATOR

An example of the circuitry adopted in a service-type signal generator is revealed in Fig. 5.2. Power is provided by a double-wound mains transformer T_1 and a bridge rectifier. The double-triode valve, V_1, is the r.f. oscillator, one section of which works on five ranges and the other section on three ranges, giving a total of eight ranges over the spectrum 100 kHz to 240 MHz on fundamentals. V_2 is the l.f. oscillator (400 Hz) which modulates either oscillator to a depth of approximately 30 per cent. VR_1 is the 'set carrier' control which works in conjunction with the r.f. monitor indicating meter M and rectifier.

F.M. SIGNAL GENERATOR

Although there are several instruments in which the modulation can be switched to either amplitude or frequency, there are also other instruments, as with

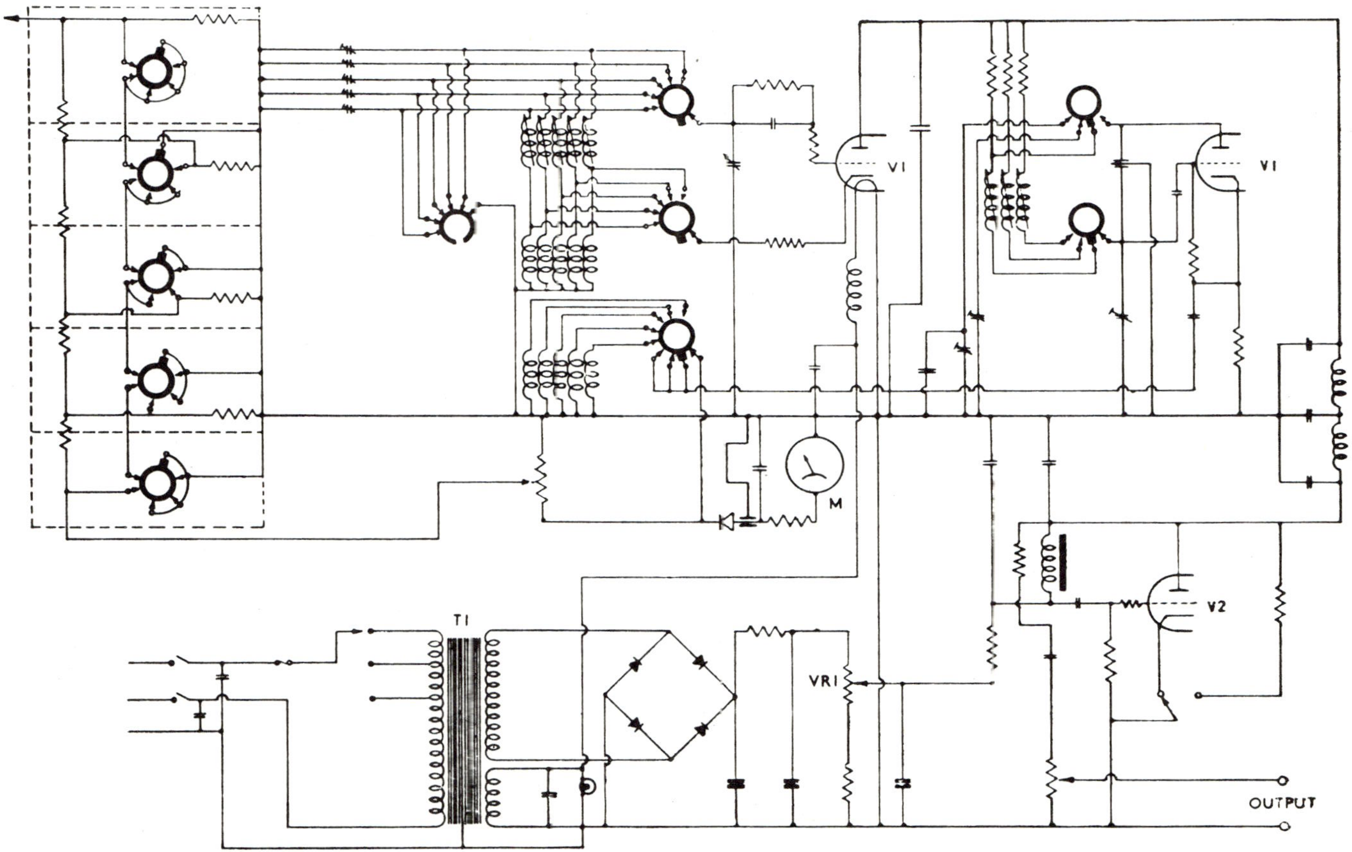

FIG. 5.2 *Circuit diagram of example A.M. Signal Generator.*

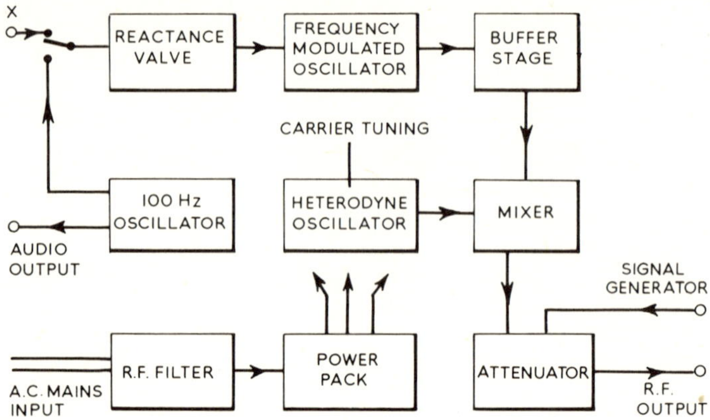

FIG. 5.3. *Block diagram of example F.M. Receiver Alignment Generator. The circuit arrangement is shown in Fig. 5.4.*

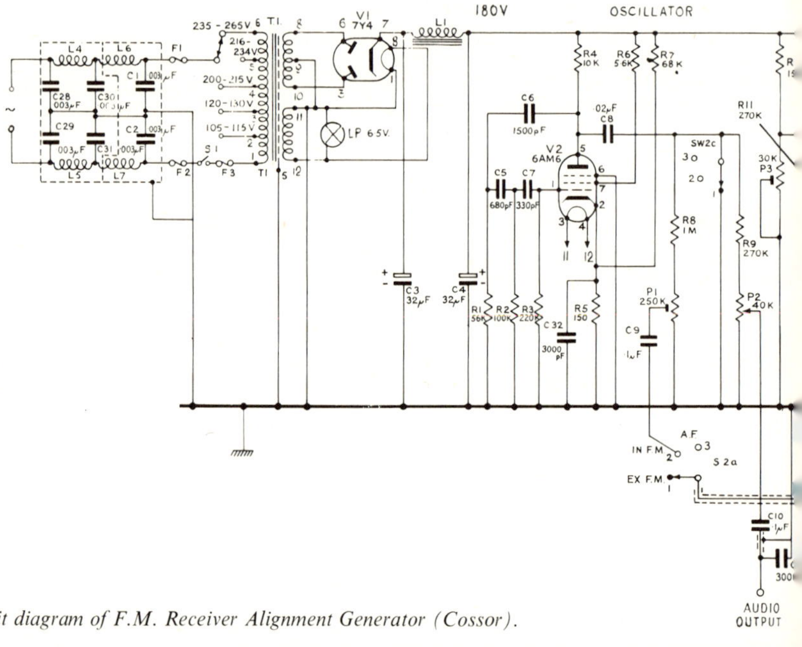

FIG. 5.4. *Circuit diagram of F.M. Receiver Alignment Generator (Cossor).*

amplitude modulation, which provide solely a frequency-modulated signal. The modulation signal is derived either internally from a 400 or 1,000 Hz a.f. oscillator or from an external source, such as the timebase of an oscilloscope, when the instrument may be used in conjunction with an oscilloscope to provide a display of the i.f., signal-frequency and discriminator characteristics of f.m. receivers.

Operating in this mode, such instruments are usually called 'wobbulators', the application of which is dealt with in a later chapter. However, at this stage it should be noted that for 'sweep alignment', as the wobbulator alignment technique is known, the modulation signal may be obtained either from the oscilloscope's timebase, as mentioned, or from the 50 or 60 Hz power frequency. Some instruments, in fact, do not have provision for the application of a timebase signal, but more about that later.

A block diagram of such an instrument is given in Fig. 5.3 and the circuit diagram in Fig. 5.4. From the block diagram it may be seen that the r.f. output is obtained by mixing the voltages from two oscillators, their frequency difference providing the i.f. coverage and their sum the r.f. coverage.

Frequency modulation is provided by a reactance valve in association with the fixed-tuned oscillator, or f.m. oscillator, whose centre frequency is set at 57·5 MHz. In the reactance valve, V_3, the r.f. in the grid circuit leads the anode voltage by almost 90 deg. The grid-cathode voltage therefore leads the anode voltage by the same angle and results in the anode current leading the anode voltage likewise. The stage thus behaves as a capacitance, whose magnitude is determined by the d.c. voltage at V_3 grid. The a.c. voltage either from the 1,000 Hz oscillator or the timebase signal of an oscilloscope is superimposed

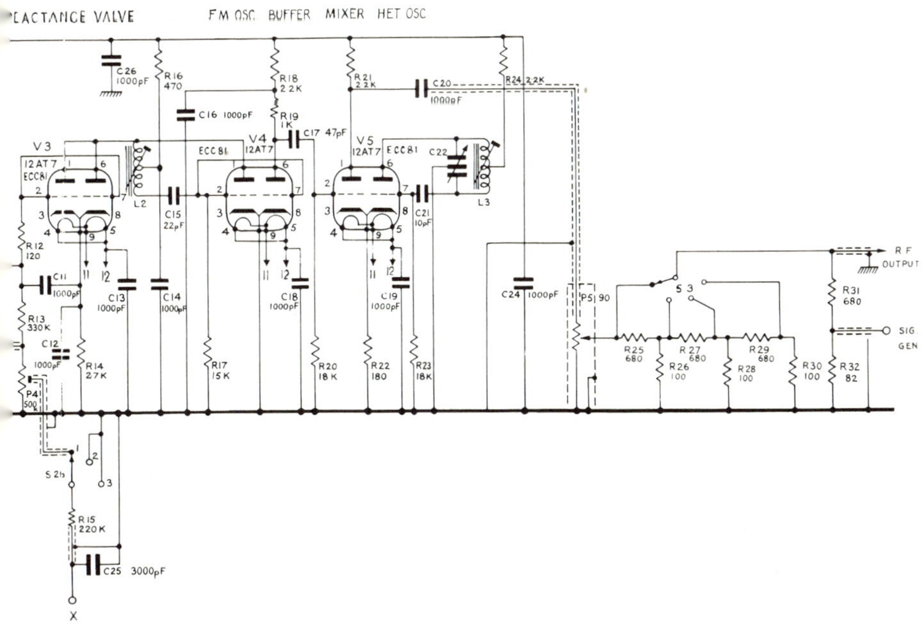

upon the d.c., thereby providing cyclic changes in the effective capacitance of V_3.

As the frequency of the f.m. oscillator is governed by the capacitive effect of V_3, it follows that the frequency of this stage is caused to deviate on either side of its fixed frequency by reason of the a.f. voltage, and in this way frequency modulation is accomplished. The remainder of the circuit can be studied from Figs. 5.3 and 5.4.

The deviation is ± 75 kHz from the internal 1,000 Hz oscillator or from ± 25 to ± 400 kHz using the oscilloscope's timebase voltage, when the instrument is used for visual displays. The r.f. output is from 5 microvolts to 50 millivolts, via the output attenuator.

FERRITE MODULATOR

In order to observe the response curve of any tuned circuit on an oscilloscope by the use of a wobbulator, the frequency deviation must be at least equal to the curve's bandwidth. Some instruments give a maximum deviation of 400 kHz, which is ideal for f.m. receivers, but, of course, is unsuitable for the wide response curves of vision channels of television sets. The reactance valve is somewhat limited in this respect, particularly where a very wide linear sweep is called for.

Another method of securing frequency modulation over a greater sweep is by the use of the so-called 'ferrite modulator'. This consists of a small ferrite core, which carries an r.f. winding forming part of the oscillator tuned circuit. The ferrite core is placed in the gap of a different type of core, which carries the sweep voltage or modulation signal winding. The permeability of the ferrite core is set to a certain nominal value by means of a small permanent magnet.

When the sweep voltage or modulation signal is applied to the appropriate winding the permeability of the ferrite core is caused to vary linearly about its nominal value as set by the magnet, and because the inductance of the r.f. winding is directly proportional to the core permeability, the oscillator frequency is caused to vary in accordance with the modulation or sweep signal, thus giving frequency modulation.

VARACTOR DIODE

More recent instruments employ a varactor diode for providing the frequency sweep. This is merely a p-n junction diode designed specifically to exploit the capacitance present across the junction when a reverse bias is applied. It will be recalled that as the reverse bias is increased the depletion zone widens, and since the depletion zone constitutes the dielectric of the capacitance (with the p and n semiconductors as the plates) it follows that the capacitance value diminishes as the reverse bias is increased.

It thus becomes possible to arrange the sweep signal to alter the reverse bias of the diode which, when connected as a capacitive element of the oscillator tuned circuit, alters the generator frequency over a predetermined range during each forward trace of the oscilloscope, returning it to the starting frequency at the conclusion of each retrace. Readers interested in the design and construction of a transistorised television wobbulator based on the varactor diode should refer to the August, September and October, 1970 issues of *Wireless World* and to the articles on the subject by W. T. Cocking.

AUDIO FREQUENCY GENERATOR

As already mentioned in Chapter 1, audio frequency generators, or oscillators as they are sometimes called, commonly adopt a Wien-bridge configuration followed by a low distortion amplifier (see the circuit of Model Si 453 by J. E. Sugden in Fig. 1.7). Low distortion sine-wave signals over the audio spectrum and above can be achieved by a well-designed circuit of this kind, and low distortion is one of the prime parameters required for distortion measurements in hi-fi amplifiers.

Another Wien-bridge circuit this time by Mullard is given in Fig. 5.5. The circuit is so called because feedback occurs via a resistor-capacitor network which provides the necessary phase shift. This is the Wien network which in Fig. 5.5 consists of the two sets of switched capacitors in association with the fixed and variable resistors. Feedback is taken from the centre or 'bridge point' of the network to the base of Tr_1.

The switched capacitors provide four ranges which are 15 to 200 Hz, 150 to

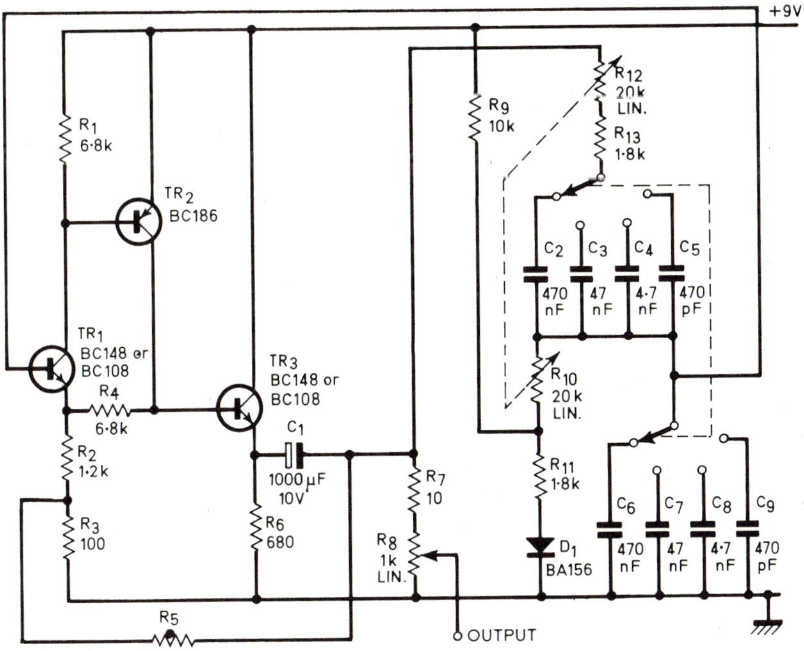

FIG. 5.5. *Circuit of Wein-bridge audio oscillator by Mullard.*

2,000 Hz, 1·5 to 20 kHz and 15 kHz to 200 kHz, the ganged variable resistors allowing the frequency to be adjusted over any selected range. As the circuit is designed the output is 1 V r.m.s. maximum, adjustable down to zero by the output control R_8 in the emitter circuit of Tr_3. The thermistor R_5 (STC thermistor type R53 or similar is suitable) controls the amplitude of the output signal and makes it independent of small changes in supply voltage and ambient temperature. The circuit will operate from a battery supply from 6 to 10 volts and consumes 6·3 mA at 6 V, 11·5 mA at 9 V or 13 mA at 10 V.

PATTERN GENERATORS

Monochrome

Pattern generators for adjusting monochrome receivers are falling out of mode, but they can still be very useful in the busy service department where it is necessary to make adjustments to receivers during normal programme periods when an 'off air' test card is not available. They are essentially signal generators modulated with video signal for yielding displays carrying some of the information of a test card. Modulation is often switchable over a number of different displays, ranging from vertical and horizontal lines, crosshatch patterns, definition bars and contrast steps. Sometimes there is a separate generator section for the sound carrier, modulated with 400 Hz or 1 kHz tone, useful for checking sound-on-vision. The best instruments produce video signal closely simulating that transmitted, with accurately processed line, field and blanking pulses.

Clearly, a pattern generator facilitates the setting-up of receivers independent of actual transmissions. It also makes it possible to locate aerial defects, interference problems and so on, while showing up faults such as misalignment, ringing, defocusing, picture non-linearity, impaired line and field synchronising, smearing, tearing, etc. Some instruments are designed so that the video signal can be used direct (e.g., without the carrier wave) for locating faults in the video and sync stages.

The r.f. signal is commonly delivered at 75 ohms via an attenuator, similar to that found on service type signal generators, from zero up to about 100 mV.

Colour

Pattern generators for colour television applications are similar to their earlier monochrome counterparts and they can, in fact, be used for the adjustment of monochrome receivers. As mentioned earlier, there is the type that produces colour pattern displays in addition to crosshatch and dot patterns, contrast steps and an unmodulated, synchronised raster, and the less costly type that provides dot patterns for static convergence, crosshatch patterns for dynamic convergence, contrast steps for grey-scale tracking and a blank raster for purity adjustments.

The colour pattern generator is an essential instrument for colour television servicing, and the minimum requirement should be for convergence patterns and displays for grey-scale tracking and purity adjustment. As this is the kind of instrument that would be employed extensively for field servicing and adjustments to receivers during the initial setting-up it should obviously be as robust as possible and not too large.

The fully colour-encoded type is effectively a colour transmitter. Elaborate models generate composite video signal corresponding to the standard colour bars, and this type, being much more expensive than the non-colour type, would probably have greatest application in the service workshop. Video corresponding to the colour information is highly desirable for checking the performance of the chroma and decoder sections of the receiver. Indeed, the instruction booklets issued with some of the costly models contain very detailed information regarding tests and adjustments of this kind that can be made to the chroma and decoder sections of receivers in conjunction with the signals generated. In some cases it is possible to appraise the colour performance and to process the adjustments merely by observing the colour displays on the screen. For more complex tests, however, the signals in the receiver need to be viewed on an oscilloscope, which is another essential instrument for colour receiver servicing.

For the adjustments made from a pattern generator to hold good on an off-air transmission the parameters of the signal produced by the pattern generator must be accurate. For example, the master oscillator must be stable and preferably energised from a stabilised voltage supply. In some instruments the master oscillator runs at twice the line frequency, being divided down to yield the line and field sync pulses. Thus a drift in master oscillator frequency will cause a drift in the frequency of the sync pulses, meaning that the receiver's timebases will lock on to the wrong frequency and thus run slightly slow or fast. In turn this means that the convergence waveforms will be in error since these are obtained from the receiver's line and field signals. If the receiver is converged on such a signal, therefore, significant errors would be present when the receiver is run from an off-air signal!

It is essential, too, for the generated sync pulses to simulate accurately those on a real transmission. If there is any deviation the line synchronising in the set, especially if of the flywheel-controlled type, as it commonly is in colour sets, could take several lines of scan to stabilise, the symptom of which is a kink at the top of the display and, in bad cases, kinks after every horizontal line of the crosshatch pattern. Such a multilated display has little merit for convergence adjustments.

Early colour pattern generators were designed for dual-standard working, but since the dual-standard colour receiver will decline rapidly over the next few years in favour of the single-standard 625-line-only version many generators are now being made for single-standard working only. Like the receivers themselves, there is a great deal in favour of the single-standard generator. It is less complicated than the dual-standard counterpart and hence less costly; there is less in it and thus less to go wrong; and since the design is optimised for the one standard only the electronics feature fewer compromises. Nevertheless, the dual-standard model is still necessary to optimise the convergence on the 405-line standard in areas where the three television channels are not yet available in the local u.h.f. group of transmitters.

The dual-standard model, of course, should cover a v.h.f. band (usually Band III) as well as one or both u.h.f. bands. Normally the single-channel model need cover only u.h.f., but in some instances it is desirable for the 625-line modulation to be carried on a v.h.f. channel since on some television relay and communal aerial systems the 625-line u.h.f. transmissions are converted down to v.h.f. channels, the receivers used on the systems thus employing v.h.f. instead of u.h.f. tuners.

Elaborate instruments producing colour as well as monochrome patterns employ 50 or more transistors to provide all the functions and the accurately

(a) (b) (c) (d)

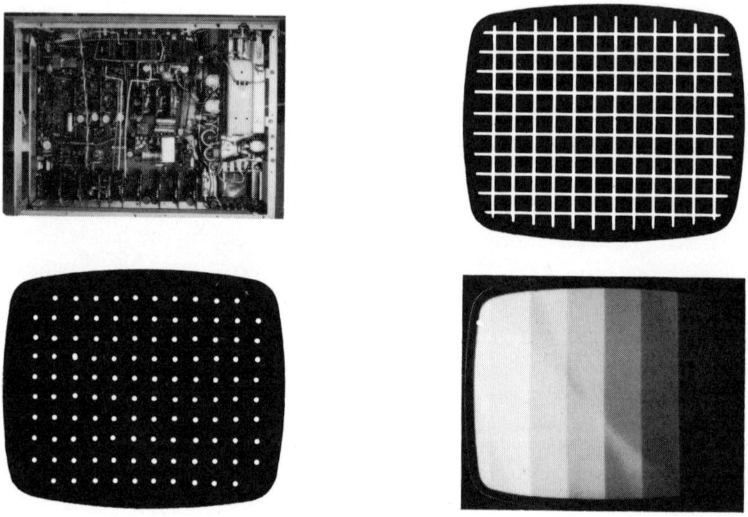

FIG. 5.6. *Internal view of Körting colour pattern generator* (a). *Crosshatch and dot patterns* (b) *and* (c). *Contrast steps in monochrome resulting from the standard colour bars, such as produced by the Körting generator* (d).

simulated components of the composite signal. Module-type printed circuit boards which plug into a main matrix are employed in the Körting Model 82512 colour generator, and some idea of the complexity of the instrument can be gained from Fig. 5.6*a*, which shows an inside view. Typical cross-hatch and dot pattern displays are shown at *b* and *c*, while *d* shows the contrast steps resulting from the standard colour bars displayed in monochrome.

ALIGNMENT OF RADIO RECEIVERS

The signal generator's prime function lies in the provision of signals suitable for the alignment of the tuned circuits of radio and television receivers. This section and some which follow deal with the problems involved, firstly the alignment procedures appropriate to the type of receiver whose circuit is given in Fig. 5.7.

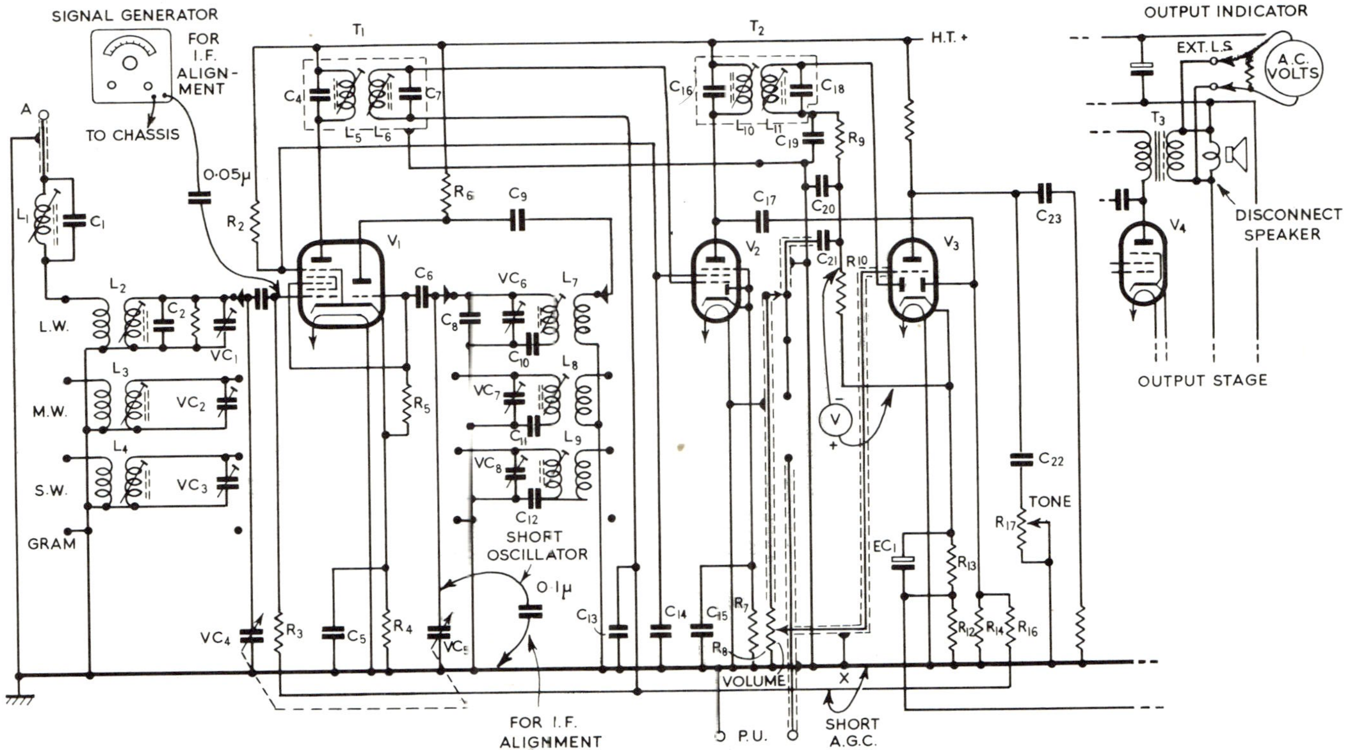

FIG. 5.7. *Circuit sections of a radio receiver showing how instruments are connected for alignment, as described in the text.*

Although this is not a very recent set, using valves instead of transistors, it is nevertheless admirably styled clearly to reveal the tuned circuits and other areas which need to come under attention during the alignment exercise.

Before alignment is undertaken, the instruments to be used, and the receiver, should be allowed to warm up adequately, so that frequency drift is minimised during the alignment process. Ten to fifteen minutes is generally sufficient time for most equipment. Also check that the tuning pointer or cursor and drum are correctly adjusted in relation to the position of the moving vanes of the tuning gang. When the pointer is at its lowest wavelength (highest frequency) position, the vanes of the gang should be fully open. Take the precautions already detailed in relation to a.c./d.c. receivers.

Intermediate-Frequency Alignment

Before the r.f. or oscillator stages can be aligned, the i.f. stages must first be attended to. This requires a modulated signal generator which tunes over the i.f. of the set under adjustment. Lists of receiver i.f.'s are usually available through trade sources.

In addition to the generator some means of detecting output is also required. This can take the form of an ordinary output meter or a.c. voltmeter connected across the loudspeaker transformer. The speaker can remain in circuit, but as the tone from the signal generator's modulation is somewhat disturbing it is often best to disconnect it and connect in place a wire-wound resistor of the same resistance as the loudspeaker's impedance at 400 Hz. On most sets a 5-ohm resistor is suitable. In the circuit of Fig. 5.7, the output meter would be connected across T_3 secondary (i.e., the 'Ext.L.S.' sockets), in parallel with the load resistor (such a load is sometimes fitted in the output meter, in which case an external load is not necessary), and the speaker would be disconnected.

High-resistance Voltmeter as Indicator

If an a.c. voltmeter or output meter is not available, the rectified i.f. signal appearing across the detector load resistor can be metered on a high-resistance voltmeter or an electronic test meter. Thus, as the tuned circuits are brought into alignment so the i.f. signal at the detector diode increases and the d.c. voltage across the load does likewise. When using this method the signal generator need not have modulation facilities, and an ordinary calibrated oscillator or frequency meter will serve. The detector load in the receiver circuit is R_{10}, across which the voltmeter is shown connected.

Short-Circuit Local Oscillator

To avoid misleading readings, whenever aligning i.f. stages, the local oscillator should be put out of action. The best way to do this is to connect a large value capacitor across the oscillator section of the tuning gang, as shown in the circuit. Where one is sure that there is no d.c. voltage between the plates of the gang, a direct short-circuit can be applied. Finally, the tuning gang should be set to the highest wavelength (*L.W.*); e.g., tuning gang fully closed, and the a.g.c. line should be short-circuited, as shown at point *X* on the circuit.

The output leads from the signal generator should be connected to the control grid of the frequency-changer valve (pin 2 of V_1), via an 0·05-microfarad capacitor, and to the receiver chassis—the screened outer or 'earthy' lead.

Tune the generator to the correct intermediate frequency; switch on modulation, if required, and adjust the attenuator to give a reasonable reading on the output indicator. If the i.f. stages are seriously out of alignment, full signal or the 'force' signal may have to be applied in order to secure an indication of output. For both types of indication described, the meter may not have to be set any higher than 1 volt full-scale.

Adjust the trimmers or dust cores in the i.f. transformers for maximum output in the following order: L_{11}, L_{10}, L_6 and L_5 (Fig. 5.7), which correspond to the secondary of the second i.f. transformer, the primary of the second i.f. transformer, the secondary of the first and finally the primary of the first.

The signal output from the generator must be kept to a minimum at all times during the alignment process, and will need to be progressively reduced as the various circuits are brought into alignment. If too great a signal is applied the stages will overload and possibly go into grid current, giving a false alignment.

Wavetrap Adjustment

Many sets are fitted with a parallel-tuned rejector (wavetrap) in the aerial circuit to avoid breakthrough of transmission on or near the i.f. This should be adjusted at exactly the same generator setting as was used to align the i.f. transformers. A very strong signal should be applied to the aerial socket (and earth) via the dummy aerial, and the rejector trimmer or core (L_1, Fig. 5.7) should be adjusted for *minimum* output.

Radio-Frequency Alignment

Apply the generator signal across the aerial and earth sockets of the set through a dummy aerial, ensuring that the 'earthy' lead (usually screened or marked 'earth' on the dummy aerial) is joined to the earth socket. Tune the set and generator to corresponding wavelengths at the high-frequency end of the *L.W.* band. Turn the volume control to maximum if the output indication is by an output meter or a.c. voltmeter connected across the loudspeaker transformer.

Adjust the *L.W.* oscillator trimmer for maximum output (VC_6, Fig. 5.7). Reset generator output level to avoid overloading and readjust oscillator trimmer.

Adjust the *L.W.* aerial trimmer (VC_1) for maximum output. Adjust the *L.W.*

trimmer of the radio-frequency amplifier (where such a stage is provided) in a similar manner.

Tune the set and generator to corresponding wavelengths at the low-frequency end of the band and adjust the core in the *L.W.* oscillator coil (L_7) for maximum output. Adjust the core in the *L.W.* aerial coil (L_2) for maximum output. Adjust the core of the long-wave r.f. coil (if fitted) in a similar manner.

The r.f. alignment of the medium- and short-wavebands should be tackled in exactly the same way, using two spot frequencies, one at the high-frequency end of the band (trimmer adjustment) and one at the low-frequency end of the band (core adjustment). Sometimes frequency calibration points are marked either on the tuning scale or tuning drum, depending on the design of the set.

In some sets cores may not be fitted in the coils, but in place there may be relatively large-value trimmers connected in series with the tuned circuits. These are known as 'padding' capacitors, and should be adjusted as described for the cores.

The adjustments at the lower frequencies will probably affect those carried out earlier at the higher frequencies; therefore, all adjustments should be repeated at both alignment points until no further improvement in tracking can be obtained.

It is obvious that maximum sensitivity will occur at those points at which alignment was undertaken, for it is virtually impossible to secure 100 per cent tracking in terms of both frequency and sensitivity over the entire band, especially where the frequency ratio is high. As the programmes on 208 metres are very popular, it is sometimes desirable to select that wavelength as the high-frequency alignment point for the medium waveband.

Image Rejector Adjustment

Some receivers feature a series-tuned wavetrap connected across the *L.W.* input transformer tuned to the frequency at which the 'image' of a powerful B.B.C. transmitter may fall. For example, the 247-metre B.B.C. Radio 1 station might well appear also at approximately 1,056 metres, due to the 'image' or 'second channel'. This could cause interference on another required station at that frequency, but can be eliminated by the wavetrap mentioned.

The rejector should be adjusted, after the *L.W.* alignment has been completed, in the following manner. Tune the generator to 247 metres and apply the signal at high level to the aerial and earth sockets through a dummy aerial. Tune the receiver on the *L.W.* band, in proximity to 1,056 metres (on receivers with a 465-kHz i.f.) until the signal is heard or indicated on the output indicator. Adjust the trimmer or core of the image rejector for minimum output.

It should be noted that after i.f. alignment of any receiver the short-circuit on the oscillator should be removed before alignment of the r.f. stages and wavetraps is attempted. The short-circuit on the a.g.c. line should remain, however, until the whole alignment is completed.

TRANSISTOR RECEIVERS

Transistor receivers are no more difficult to align than their valve counterparts, and merely because transistors instead of valves are used there is no difference in the basic procedures from those already described.

Since most transistor sets incorporate a ferrite rod aerial and have no aerial socket as the valve set in Fig. 5.7, some scheme has to be devised to inject signal into the aerial circuit. The best way is to wind several turns of insulated copper wire on a small former and to connect the ends to the low impedance signal output of the generator. The former can then be placed near to the ferrite rod aerial. Some transistor portables have a socket for connecting a car-type aerial. It is possible to inject the signal in at this point from the 'live' signal lead, connecting the 'earthy' side to receiver chassis or the 'earthy' side of the battery.

A circuit of Mullard design showing the frequency changer, i.f. and detector stages of a transistor receiver is given in Fig. 5.8. This shows that the ferrite rod aerial T_1 also constitutes the aerial tuning, the coupling to the base of the frequency changer transistor TR_1 being by L_2, with L_1 the tuned winding. The local oscillator transformer consists of windings L_3, L_4 and L_5, and C_1 section of a two-gang capacitor tunes the aerial and C_5 section the local oscillator.

The i.f. transformers are T_3, T_4 and T_5, tuned by parallel capacitance and dust-iron cores. A.G.C. bias is taken from the anode circuit of the detector diode D_1 (OA90) to the base circuit of TR_2, which is the first i.f. amplifier. TR_1 is arranged as a self-oscillating mixer (the function being that of frequency changing). TR_3 is the second i.f. amplifier which feeds the detector diode via T_5.

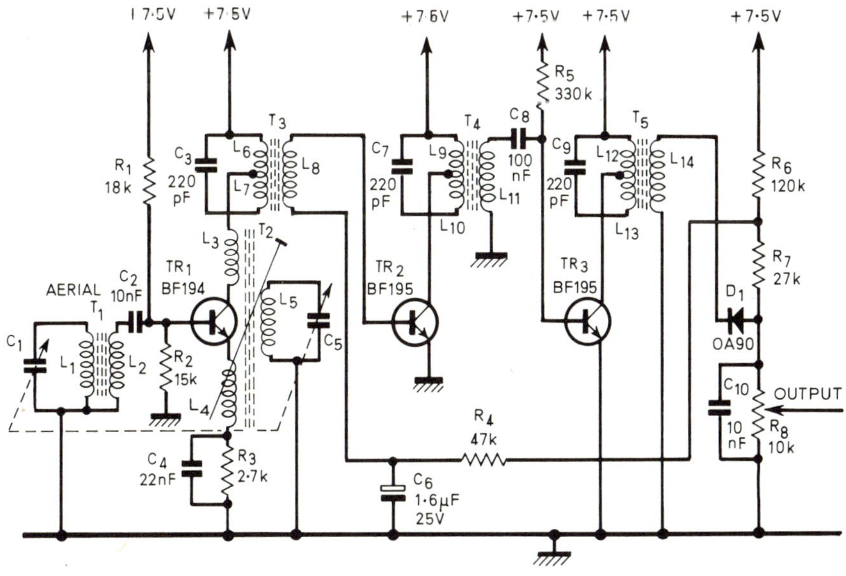

FIG. 5.8. *Circuit of the r.f. and i.f. stages of an a.m. receiver by Mullard. See text for alignment details.*

A small forward bias (15 μA) is applied to the detector diode from the potential-divider consisting of R_6/R_7 in the top arm and R_8 in the bottom arm. This enhances the small-signal efficiency and keeps the loading on T_5 constant.

The i.f. transformers are aligned by adjusting the dust-iron cores, as also is the oscillator transformer. For 'trimming' at the h.f. end of the band some sets use small preset capacitors across the gang sections, the 'padding' at the l.f. end then being handled by adjusting the cores of the r.f. (when used) and oscillator coils.

The Mullard circuit uses Lockfit transistors which are supplied to set makers in sets of three. TR_1 operates with a V_{ce} of 4·8 V and an I_e of 1 mA, and voltage biasing serves to stabilise the operating point against h_{FE} and V_{be} spreads. R_3 and C_4 control and stabilise the oscillator signal.

All transistors are npn type, so the supply needs to be positive with respect to the emitters. The i.f. stages are designed to handle large signals without overloading, and the a.g.c. provides a signal range of about 45 dB for a 6 dB change in output.

Care has to be taken with transistor sets in connecting an output meter. Some models do not favour running without a speaker load, while a short-circuit across the speaker under conditions of drive can ruin the output transistors and possibly the driver. However, an indication can be obtained by substituting a resistor in place of the speaker and using an audio mullivoltmeter to monitor the modulation signal.

The circuit in Fig. 5.8 is for medium-wave only and this would be very easy to align. The plan would be to inject an i.f. signal (via a capacitor of about 0·01 μF) to the base of TR_1, mute the local oscillator by putting a short across C_5 and then adjust the i.f. transformers for maximum output at the detector or at the output of the power amplifier section (not shown in Fig. 5.8). The standard a.m. i.f. is 470 kHz.

After this the oscillator short would be removed and a medium-wave signal (modulated, of course) coupled to the ferrite rod. The top end of the band would normally first be aligned to the tuning scale by adjusting the oscillator trimmer (as mentioned, usually across the oscillator section of the gang). The bottom end of the band would be likewise adjusted but this time by the oscillator core (or, sometimes, by a 'padding' capacitor).

Finally, the sensitivity would be maximised at the top of the band by the aerial gang trimmer (when fitted) and at the bottom end of the band by any 'padding' components. The alignment instructions sometimes specify the sliding of the aerial coil along the ferrite rod to secure maximum sensitivity.

Other wavebands when fitted are adjusted in a similar manner, trimming at the high-frequency end of the band and 'padding' at the low-frequency end, and details of this have been given in connection with the valve receiver in Fig. 5.7.

It is not so easy to mute the a.g.c. system in transistor sets as it is in valve models for, like that in Fig. 5.8, the a.g.c. is commonly a part of the controlled transistor base circuit. The best plan is to align with the a.g.c. active and to ensure the least error by keeping the input signal at the lowest possible amplitude.

SPOT-FREQUENCY ALIGNMENT OF F.M. RECEIVERS

As is common with a.m. receivers, the alignment process starts at the i.f. stages. Also, some means of indicating output is required. This can be an output meter or a.c. voltmeter connected in place of the loudspeaker, as already described but in this case, in order to secure an audio output, the signal source will need to be frequency-modulated. However, the alignment process will be exactly the same (up to the detector) as that described later.

Possibly a better way of securing an output indication is by connecting a high-resistance voltmeter or a valve-voltmeter across the load of the ratio detector. This will record the rectified i.f. signal and does not require a modulated signal source. Thus, an ordinary a.m. generator with the modulation switched off will suffice. Where a balanced ratio detector is featured, the voltmeter should be connected between chassis and the audio take-off point (positive to chassis).

It may be necessary to de-tune the secondary of the ratio detector transformer by unscrewing its core until it is well away from the winding. This causes unbalance which, in certain sets, aids in the production of an indication of resonance in relation to adjustments to the i.f. transformers. The 'live' r.f. lead of the signal generator (tuned to the i.f. frequency) should be connected to the control grid of the final i.f. amplifier valve—that is, the valve immediately in front of the ratio detector and the screened lead should be connected to chassis (dummy aerial not used). After the set has warmed up, the core in the *primary* of the ratio detector transformer should be adjusted for maximum reading on the voltmeter.

Quite a large signal will need to be applied as only one i.f. stage is in use at this time, but the voltage reading is usually fairly high on a 20,000 ohms-per-volt meter. The signal should now be transferred to the control grid of the next i.f. valve in line, going backwards towards the beginning of the set, and the cores in the associated i.f. transformer (but not the one in the grid circuit of the valve to which the signal is applied) should be adjusted for maximum reading, adjusting first the primary and then the secondary cores.

This process should be continued until the first i.f. transformer in line has received attention. In order to adjust the first i.f. transformer—the one with its primary in the anode circuit of the mixer valve—the signal will have to be applied to the control grid of the mixer valve. It is usually recommended that the generator signal be loosely coupled to this circuit, and on commercial sets, including a.m./f.m. models, there is usually a point on the tuner unit to allow connexion of the signal without undue disturbance to associated circuits.

A successful method of injection is by winding a length of wire around the envelope of the mixer valve, the end of which is connected to the 'live' r.f. lead of the generator. On sets which use a screening can on the mixer or frequency-changer valve, signal connexion can be made to the can after it has been slid up the valve to disconnect it from chassis. The gain of the i.f. section at this point will be very high, and is the reason why it is possible to couple the signal capacitively, as described. The signal input should be reduced accordingly, and this applies also as the signal is progressively transferred from grid to grid along the i.f. channel.

Alignment of the ratio detector should next be undertaken. This is an important exercise, for unless an accurate balance is achieved, not only will the reproduction be below standard, but amplitude interference may also be unduly troublesome.

In the case of an unbalanced ratio detector, the voltmeter should be transferred from across the load to a point of balance in the circuit. Balance is synthetically created by interposing a pair of 47,000-ohm matched resistors in the circuit, as shown by the broken lines in Fig. 5.9. The meter is then connected between the audio take-off point and the junction of the two resistors. In the balanced ratio detector it not usually necessary to remove the meter from its original point, as the circuit was purposely put out of balance to start with by de-tuning the ratio detector transformer.

If a high-resistance meter or valve-voltmeter is not available, a 50-microampere meter connected in series with a 100,000-ohm resistor or a 100-microampere meter connected in series with a 47,000-ohm resistor can be used

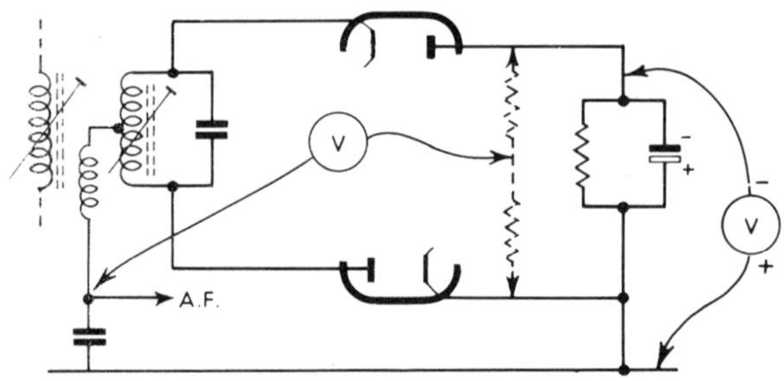

FIG. 5.9. *Showing how an indicating voltmeter should be connected when aligning a ratio detector.*

instead. An unmodulated i.f. signal is then applied to the front of the aligned i.f. channel, and the signal level adjusted to provide a reasonable indication on the the meter. Balance is secured by carefully adjusting the core in the *secondary* of the ratio detector transformer for *zero* reading on the meter. This will occur midway between a negative-going and positive-going deflection. Some valve-voltmeters have centre-zero provision to facilitate this operation.

Foster-Seeley Discriminator Alignment

Certain high-quality f.m. tuners (mostly Austrian) use this type of detector, alignment of which can be undertaken by the use of an a.m. generator and output meter or a.c. voltmeter.

The discriminator should be unbalanced by unscrewing the core in the secondary winding of the discriminator transformer. An amplitude-modulated signal at the correct i.f., should be applied as already described, and this recorded on an output meter or a.c. voltmeter connected in place of the loudspeaker. After normal adjustment of the i.f. stages, and without altering the generator frequency, the core in the secondary winding of the discriminator transformer should be adjusted for *minimum* reading on the output meter, if necessary increasing the signal level to give an adequate reading.

R.F. Alignment

A frequency-modulated signal should be applied to the receiver aerial socket, direct from the generator. On most sets it is required to align at one spot frequency at about 93 MHz; a calibration mark is provided on the scale for this purpose. The set and generator should be adjusted to corresponding frequencies and, with an output meter or a.c. voltmeter connected to the set, the oscillator trimmer should be adjusted for maximum output. The aerial and r.f. trimmers are then adjusted for maximum output, reducing the signal in order to maintain the smallest possible deflection on the meter with the volume control at maximum.

SPOT-FREQUENCY ALIGNMENT OF TELEVISION RECEIVERS

In addition to providing an output indication in the sound channel, when aligning television receivers a similar kind of indication must be made available in the vision channel. One of the best methods of securing a vision indication from a monochrome set is by connecting a d.c. voltmeter between the cathode of the picture tube and the chassis of the receiver, as shown in Fig. 5.10*a*. An increase in vision signal will result in a decrease in voltage at the cathode, but there will always be a standing voltage of between 50 and 150 volts, depending on the design of the set.

Another method is shown in Fig. 5.10*b*. Here the current in the anode circuit of the video amplifier valve is measured by interposing a milliampere meter at the h.t. side of the anode load resistor. An increase in signal, due to the i.f. and r.f. circuits being brought into alignment, for instance, will cause an increase in current.

Alternatively, the voltage across the anode load resistor of the video amplifier valve may be measured. This again will rise with increase in signal, and is shown in Fig. 5.10*c*. In all cases, the meter connecting leads must be kept as short as possible to avoid instability and incorrect readings due to positive or negative feedback. In the case of Fig. 5.10*b*, it sometimes pay to connect an 0·1-microfarad capacitor between the top of the load resistor and chassis, for decoupling.

It is possible to get a video indication at the cathode points of the picture tube of colour sets when colour-difference drive is employed at the tube grids, but it is highly desirable to adopt the visual method of aligning colour receivers in which the Y deflection of the trace on the cathode-ray tube screen indicates the extent of the response (see Chapter 6).

For indicating the sound signal, the same methods may be employed as detailed for alignment of sound receivers. As mentioned earlier, due precautions should be taken when dealing with a.c./d.c. equipment, regarding instrument isolation and so on. Precautions should also be taken against personal injury when dealing with the inside of television receivers which need to be switched on and connected to the power supply.

The test signal from the generator is applied in the first instance to the control grid of the frequency-changer valve, via an 0·05-microfarad capacitor ('live' lead) and the screen to chassis (dummy aerial not required). The 'earthy' lead should also be isolated by a capacitor of similar value on a.c./d.c. chassis. As in broadcast receiver practice, the i.f. stages are the first to be adjusted and, as the frequency-changer is common to both sound and vision, there is no need to remove the signal from this point during the whole process of aligning both sound and vision channels and the sound rejectors.

If the set is considerably out of alignment, a strong signal will first have to be applied to secure a workable indication on the output indicating meters. As each individual circuit is brought into alignment, however, the signal output must be reduced to keep the indicated output substantially constant. It is also necessary to maintain the lowest input signal, consistent with usable indication on the output meters, to prevent the a.g.c. systems from counteracting the increase in gain as the circuits approach resonance. In some circumstances, it is recommended in the makers' service manuals that the a.g.c. systems be disconnected during alignment.

The signal generator should be adjusted to the sound i.f. (unmodulated) and the cores of the sound rejectors adjusted for *minimum* response on the vision output indicator. Remember that the arrangement in (*a*) of Fig. 5.10 will result in a *greater* reading, while those in (*b*) and (*c*) will result in a *smaller* reading. When the rejectors are correctly aligned a trough will occur in the vision response curve at the sound i.f.

The sound i.f. stages should next be adjusted at exactly the same generator frequency, but with the modulation switched on. The cores in the i.f. transformers, starting with the final sound i.f. secondary and working towards the front of the circuit, should be carefully adjusted for maximum indication on the sound output meter. The adjustments should be repeated several times until no further improvement in sensitivity can be obtained.

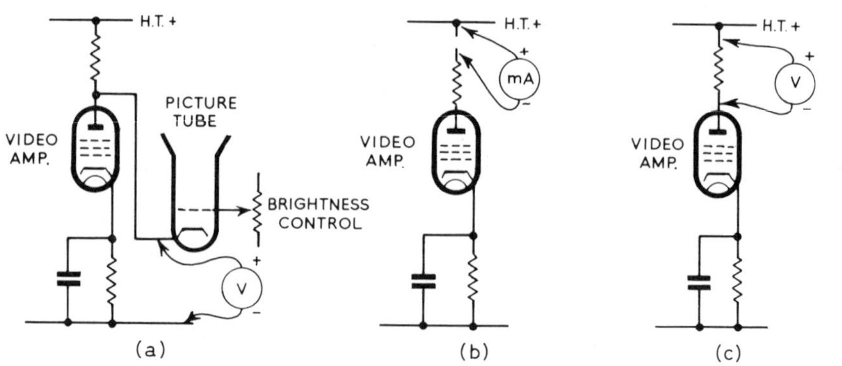

FIG. 5.10. *Methods of connecting an instrument for indicating vision output. The meter leads must be as short as possible to avoid instability and incorrect readings due to feedback.*

For aligning the intercarrier sound channel of dual-standard and 625-line-only (single-standard) receivers the generator frequency must be accurately adjusted to 6 MHz, established either by beating the generator signal with the intercarrier signal or by a crystal check facility. The intercarrier channel feeds the f.m. detector, so the alignment in this area of the receiver is handled in a manner similar to that explained under the section dealing with the alignment of f.m. receivers.

No hard-and-fast rules can be given for the vision i.f. stages, as the alignment procedure depends very much on the design of the circuits. Where possible, the maker's service data should be acquired. It is often necessary, in order to prevent interaction when adjusting the primary of a vision i.f. transformer, to damp the secondary with a 470-ohm carbon resistor and conversely, when adjusting the secondary, to connect the damping resistor across the primary. Such information is given in the service manual, but usually applies only to receivers with over-coupled transformers.

The signal can generally be kept connected as for sound i.f. alignment, but the frequency should be altered to correspond to the various tuning points of the bandpass filters and rejectors. Recent single-standard 625-line-only receivers are composed in the i.f. channel of two sections, one dealing with the bandpass characteristics and the other with the various rejectors, including that corresponding to the sound i.f., which is often designed to put a dip of about 35 to 40 dB in the response characteristic. The intercarrier sound signal at 6 MHz is obtained at the vision detector, it representing the difference between the sound and vision carriers. Unless the level of the sound i.f. signal is correct the intercarrier signal can suffer noise and picture signal components, resulting in the symptom known as intercarrier buzz.

The oscillator trimmer or core corresponding to any v.h.f. channel may, however, be adjusted to provide maximum output on the sound carrier, consistent with minimum sound-on-vision, but this is best adjusted on an actual transmission. Before making this adjustment, the fine tuning control should be set to the centre of its range.

If an aerial trap is fitted on the tuner, this should be adjusted for *minimum* vision response when a high-level signal equal to the vision i.f. or other spurious response is injected at the receiver's aerial socket. The precise shapes of the sound and vision responses are moulded by the foregoing adjustments to the i.f. circuits, while the correct relationship of the vision carrier with respect to the vision i.f. response will automatically occur (provided i.f. alignment has been carried out correctly) when the oscillator is correctly tuned.

The aerial and r.f. coils in a v.h.f. tuner can, if necessary, be adjusted for maximum output on the vision meter when a signal which is approximately 1·5 MHz lower than the vision carrier is applied to the aerial socket.

CHECKING BANDWIDTH AND SENSITIVITY

A rough idea of the overall sensitivity and bandwidth of a receiver can be obtained by the use of a signal generator and a voltmeter. Sensitivity is usually expressed as the signal input required to give standard output from the receiver. The input is the e.m.f. from the generator, modulated 30 per cent at 400 Hz. The standard output depends on the type of equipment under test, and is usually stipulated in performance specifications in manufacturers' instruction and service manuals.

The American definition of standard output is 0·5 watt for receivers capable of delivering at least 1 watt of undistorted output, 0·05 watt for receivers capable of delivering 0·1 watt maximum, and 0·005 watt for receivers with less than 0·1 watt maximum undistorted output. For car radios the output is usually taken at 1 watt. The British definition is nearly always 50 milliwatts.

The output in all cases is measured across a non-inductive resistance connected in place of the loudspeaker speech-coil with a value equal to the impedance of the loudspeaker at 400 Hz. The output meter or a.c. voltmeter can be used to indicate a.f. power, as discussed in Chapters 3 and 11.

The signal generator output should be applied to the aerial socket of the receiver (through a suitable dummy aerial in the case of ordinary sets) and, with an output indicator connected as described above, the level of the modulated r.f. should be carefully adjusted to give the required output with the volume control at maximum and the tone control at minimum top-cut. The output as indicated on the generator controls is representative of the sensitivity (in microvolts). It is usual practice to record the results at three points on each waveband.

The same procedure is adopted in television sound channels, but the signal is applied direct to the aerial socket from the low-impedance r.f. output. Vision sensitivity is measured by connecting an output indicator across the video anode load (Fig. 5.10*c*). An alteration of approximately 20 volts across this resistor usually corresponds to peak-white signal level, but may vary between receivers. An *unmodulated* signal should be applied to the aerial socket at low impedance at a frequency appropriate to the vision carrier of the channel used, and the generator's attenuator adjusted until the video indication is 20 volts above the static reading with no input signal. The reading in microvolts sensitivity for peak-white picture can then be read direct in terms of microvolts from the attenuator, as in the case of the sound channel.

A graph of the overall sound and/or vision channel can be produced by plotting the alteration in output against small changes in frequency within the passband. The ideal method, however, is to view the actual response curve on the screen of an oscilloscope. How this can be done is explained in Chapter 6.

SIGNAL TRACING

Although signal generators are mostly used for the alignment of radio and television receivers, there are numerous other uses to which they can be put in association with an electronic voltmeter or oscilloscope. For example, they can be used with one or other of these instruments to trace the signal through the various r.f., i.f. and a.f. stages of a receiver.

An electronic voltmeter, suitable for reading r.f., can be connected to the anode or collector of the final i.f. amplifier valve or transistor and the signal generator applied first to the control grid or base of the same device and then to the control grids or bases of preceding devices, as a means of locating the faulty stage through which the signal will not pass.

The generator will, of course, need to be tuned to the frequency corresponding to the stage in question—to the intermediate frequency on an i.f. stage and to the radio-frequency on an r.f. stage. If no response is indicated through the final device, then it is pointless to test at preceding stages until the fault is cleared. It should be remembered that when only one stage is in circuit, a fairly high-level signal will need to be injected to secure a reading on the electronic voltmeter. Progressively smaller signals should be injected as the generator is connected through more valves.

If the signal stages are followed by a detector and a.f. stage, as in an ordinary receiver, then the applied signal can be modulated and the loudspeaker or an output meter used to indicate output.

The same procedure may be used in audio amplifiers, but here either the a.f. output from a signal generator or an audio oscillator or generator should be used. If the amplifier is 'dead', the faulty stage can quickly be located by applying the a.f. signal in turn to each control grid, starting at the output stage valves and working back towards the input stages.

CHECKING A.G.C.

To check whether or not the a.g.c. system in a radio or television set is working, an electronic voltmeter should be connected to the a.g.c. line with the meter positive to chassis and the generator signal applied to the aerial in the normal

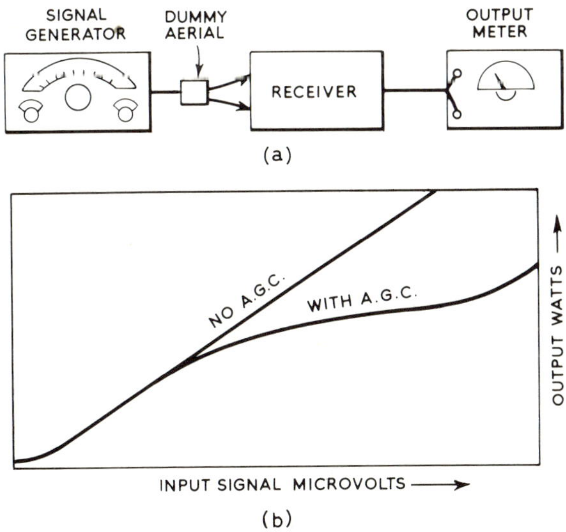

FIG. 5.11. (a) *The set-up for checking a.g.c. characteristics, using a modulated generator signal and with an output meter connected across the loudspeaker transformer.* (b) *Graph showing the curves produced when a.g.c. is working normally and when it is inoperative.*

manner. The receiver and generator should be tuned to corresponding frequencies and the signal level gradually increased while observing the meter reading. This should rise negatively with increase in signal.

Each controlled stage may be checked likewise by connecting the negative probe of the meter to the low r.f. end of the control-grid circuit of each stage in turn. If the a.g.c. line voltage rises normally negatively and one of the controlled stages (at the grid circuit) does not, then either a capacitor is short-circuited or an a.g.c. feed resistor is open-circuited.

A similar a.g.c. check may be made, using a modulated generator signal and with an output meter connected across the loudspeaker transformer. The receiver and generator should be tuned to the same frequency; the receiver's volume control should be turned full on, and the generator attenuator adjusted to give a readable indication on the output meter. The generator output should then be increased in small steps and both the signal increase and the output increase plotted in the form of a curve. If the a.g.c. system is in order, the output will rise rapidly with the first increases of generator signal, but will soon flatten out as the generator signal is increased further (see Fig. 5.11).

SECOND-CHANNEL OR IMAGE RESPONSE

Any superhet receiver, radio or television, has a response at a frequency which differs from the frequency to which the set is tuned by twice the intermediate frequency. If the local oscillator is working above the signal frequency the second-channel response will be above the tuned frequency, and below the tuned frequency if the local oscillator is working below the signal frequency.

For example, on a receiver with the oscillator above the signal frequency and an i.f. of 470 kHz, the second-channel response will fall at 1,940 kHz when the set is tuned to 1,000 kHz. If the oscillator is below the signal frequency, then the second-channel response will fall at 60 kHz. The amount of response at the second channel should be very low on a well-designed set with good pre-frequency-changer selectivity, and is usually measured in terms of a ratio.

The second-channel response may be measured as follows: Connect a modulated signal generator, via a dummy aerial if necessary, to the aerial and earth input of the set under test and tune the set and generator to the same frequency. Connect an output meter to the set and adjust the generator signal level to give a workable reading on the output meter. Make a note of the actual signal level applied to the set and the output meter reading. Also make sure that the set is accurately tuned to give full output. Next, tune the generator to the second-channel frequency and increase the generator signal level until the same reading as before is obtained on the output meter, again making sure that the set is accurately tuned to the second channel. The difference between the two signal levels applied to the set is the second-channel or image ratio, and is often expressed in decibels.

FINDING UNKNOWN INTERMEDIATE FREQUENCIES

It often happens that the technician or the experimenter is faced with the task of aligning a receiver without having knowledge of the i.f. The unknown i.f. can be found, however, by the following procedure.

Connect a modulated generator as for i.f. alignment and also an output meter or indicator. Apply a fairly strong signal and turn the receiver's volume control full on. Slowly adjust the generator tuning from about 500 kHz to 400 kHz, then to 300 kHz, then to 200 kHz and so on until a signal is indicated on the output meter. To help detect small signals, it pays in this case to keep the loudspeaker connected. Make a note of the exact frequency at which a response occurred and continue tuning the generator to see if there are other responses. If no other signals are heard, the one first noted is possibly the correct i.f.

To prove this, tune the generator to half the noted frequency, increase the signal level if necessary, and search for the harmonic. Then tune the generator

FIG. 5.12. *Avo Type HF 133 Signal Generator. This is mains operated and designed primarily for the radio and television service engineer. Frequency range is from 135 kHz to 230 MHz in six bands with a maximum output of 100 mV into a 75-ohm load. Unmodulated or amplitude-modulated signals are available, the latter being 1 kHz sine wave. This signal is also available at the l.f. output terminal.*

to twice the first noted frequency. If there is no response at that frequency, it can be assumed that the first frequency is the correct i.f. or, at least, near to the correct i.f., assuming that the dust-iron core or trimming capacitor is set to a fairly normal position in each of the i.f. transformers. For example, if the response was at, say, 108 kHz, then the i.f.s should all be aligned to 110 kHz, which is an early peaking frequency.

A further check is to set the generator at a point of suspected i.f., then with a very strong signal level re-tune the generator until the second-channel response is heard, this frequency being noted. The correct i.f. would then be half the difference between the suspected i.f. and the second-channel response.

This procedure is extremely difficult to adopt in television receivers, however, particularly in the vision i.f. stages which have staggered and over-coupled tuned circuits, and it is usually necessary to obtain some idea of the i.f.s, at least. These are given in manufacturers' service manuals.

Important alignment frequencies are the 405-line sound and vision i.f.s of about 38 MHz and 34·5 MHz respectively and the 625-line i.f. with the sound carrier falling at 33·5 MHz (well down the response) and the vision carrier at 39·5 MHz. In colour sets this puts the chroma at 1·57 MHz from the sound i.f. and about 4·43 MHz from the vision i.f., thus making the chroma i.f. about 35·07 MHz.

The application of signal generators for hi-fi adjustments and servicing is considered in Chapter 11.

CHAPTER 6

The Oscilloscope

THE OSCILLOSCOPE is an instrument which measures and displays voltages whose amplitude is changing with time. A typical continuous change in amplitude of a voltage with time is represented by the sine wave. The two components, time and amplitude, are measured, respectively, against the horizontal and vertical scales of a graticule, as shown in Fig. 6.1. Such a waveform may be readily displayed on the screen of an oscilloscope in the form of a luminous trace, the time scale being known as the X axis and the amplitude scale as the Y axis.

If the time of the changing amplitude of the voltage is of the order of several

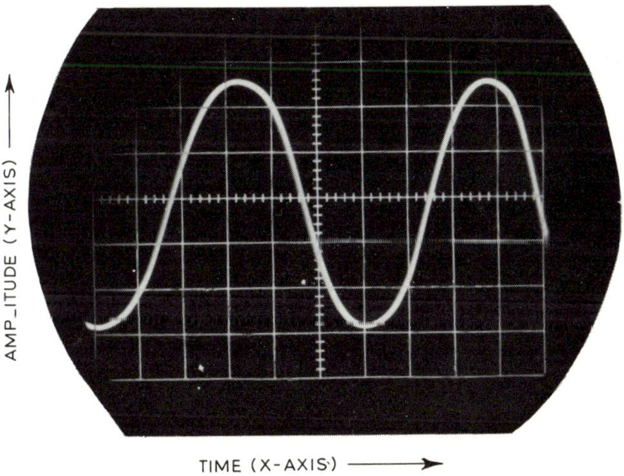

FIG. 6.1. *Showing the two components—time and amplitude—of a waveform.*

seconds, an ordinary d.c. voltmeter would register the change by the pointer falling and rising during a corresponding period of time. Under these conditions it would probably be possible to plot a graph of the waveform by using the second hand of a watch to give the time scale and instantaneous values of voltage to give the amplitude scale.

However, as the time is made smaller, the pointer would oscillate so rapidly that it would be impossible to take instantaneous readings, and as the time is further decreased the pointer would be unable to follow the rapid voltage changes and would thus become stationary. The reason for this is that the mass of the moving coil and pointer of an indicating meter is relatively large and this, coupled with friction and air-damping, makes it impossible for the pointer to follow very rapid voltage changes. Nevertheless, semi-mechanical oscilloscopes were at one time produced in conjunction with an optical system of mirrors and beams of light, but the mass of the mechanical system severely restricted its applications.

THE ELECTRON BEAM

The heart of the electronic oscilloscope is the cathode-ray tube, in which is produced a fine beam of electrons by means of an *electron gun.* This consists of an assembly of electrodes in the neck of the tube which brings the electrons to a fine focus and also accelerates them to a very high velocity. On the inner face of the tube is a fluorescent screen upon which the electron beam impinges, and which possesses the property of changing the kinetic energy of the electrons (derived by virtue of their mass and velocity) into light radiation.

Situated between the electron gun and the screen are two pairs of metal plates, called deflecting plates. These make it possible to deflect the electron beam both in horizontal and vertical directions across the screen, the extent of the deflection depending on the value of the voltage applied across associated plates. The plates which are concerned with deflecting the beam horizontally are called the X plates, while the plates which deflect the beam in a vertical direction are called the Y plates. The arrangement of the deflecting plates is shown in Fig. 6.2.

Deflection of the beam takes place because the beam, being composed of

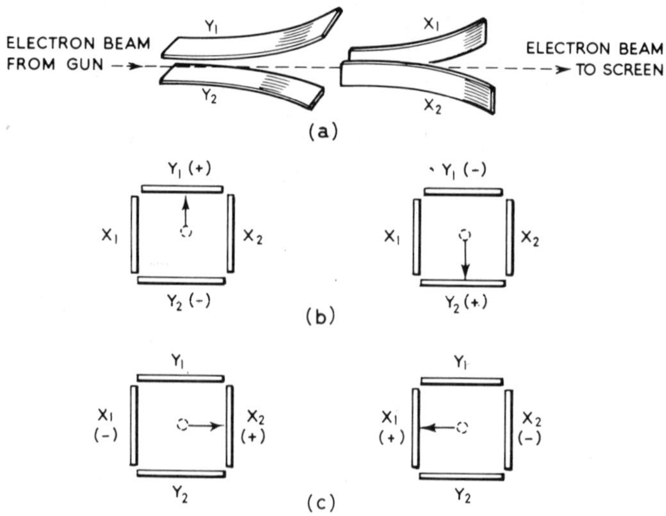

FIG. 6.2. *The arrangement of the deflecting plates of a cathode-ray tube is shown at* (a). *In the end-on views of the deflecting plates, the arrows show the direction of vertical or Y deflection* (b), *and the horizontal or X deflection* (c).

electrons, is negative and is thus attracted by a positive charge and repelled by a negative charge, and such charges are applied to the deflecting plates to provide the required horizontal and vertical deflection. The greater the amplitude of the deflecting voltage, the greater will be the potential difference between the associated pair of plates and the greater will be the deflection of the electron beam and hence the pin-point of light on the fluorescent screen. Deflection is directly proportional to the deflection voltage, so if a potential difference of 2 volts gives a deflection of 3 cm, then 4 volts will give a deflection of 6 cm. It will be realised that by applying suitable potentials to the two pairs of plates the focused beam may be deflected to any point on the screen.

The electron beam has no mechanical inertia, and for this reason can follow changes in voltage up to extremely high frequencies.

In order to produce and focus the electron beam, the gun assembly of the tube requires the application of various fixed voltages: a fairly high e.h.t. voltage

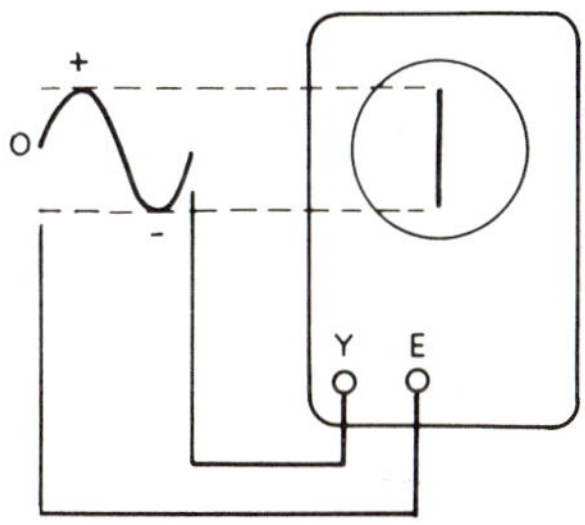

FIG. 6.3. *A vertical line corresponding in length to the peak-to-peak voltage of an a.c. waveform will be traced on the screen when the voltage is applied to the Y plates.*

(often between 1·4 kV and 4 kV) and smaller voltages for focusing the beam. If a cathode-ray tube is set up and energised in this way a sharply focused spot of light will appear in the centre of the screen. When a voltage is applied across the X plates the spot will move from the centre to the right or to the left, depending on which way round the voltage is connected, and likewise up or down if applied to the Y plates. The tube may therefore be used as a d.c. voltmeter by calibrating the deflection in terms of voltage.

If an alternating voltage is applied, up and down or left and right movements will occur, since the voltage is changing from positive through zero to negative (Fig. 6.3). If the frequency of the alternating voltage is fairly high, the eye will be unable to follow the spot movement but will discern a solid luminous line on the screen, whose total length will be equal to the peak-to-peak value of the applied voltage. Thus, the tube may also be used as an a.c. voltmeter. Half the length of the line will equal the peak value, and 0·707 times the peak value will equal the r.m.s. value.

THE TIMEBASE

For a large number of applications the electron beam is made linearly to traverse the screen in a horizontal direction, while the voltage waveform under examination is arranged to deflect the beam vertically.

A timebase built inside the oscilloscope provides horizontal deflection by subjecting the X plates to a sawtooth voltage, which rises linearly with time and moves the spot from left to right across the screen. At the finish of the rising side of the sawtooth waveform, the spot is at the right-hand side of the screen, but quickly returns to the left-hand side to recommence the horizontal trace as the sawtooth voltage on the X plates swiftly changes from maximum to minimum. The actual working traverse is when the spot is deflected from left to right, called the forward trace; the return trace or flyback occurs at the finish of the working trace, and as the speed of the flyback is many times greater than the forward trace, the spot can hardly be discerned as a luminous line owing to its very high speed across the screen. Circuits are also incorporated to blank-out the flyback when required.

Waveform Structure

Let it be supposed that a sample of the 50 Hz power voltage is applied to the Y plates and that the timebase repetition frequency is adjusted to 50 times per second. The spot will thus be deflected in both the vertical and horizontal directions and will trace a reproduction of the waveform of the actual signal applied

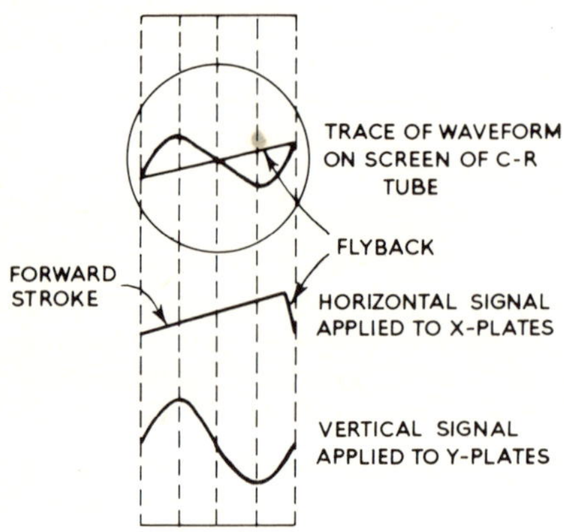

FIG. 6.4. *Showing the production of a waveform on the screen of a cathode-ray tube due to an applied Y signal and the sawtooth waveform of the timebase.*

to the Y plates, as shown in Fig. 6.4. It will be appreciated, of course, that for one complete sine wave to be traced the repetition frequency of the timebase must be equal to the frequency of the signal applied to the Y plates.

SYNCHRONISATION

If the repetition frequency of the timebase does not coincide exactly with the frequency of the applied signal, then the waveform on the screen will appear to be moving either to the right or to the left, depending on whether the timebase is running slightly faster or slower than the signal frequency. This happens because a displayed waveform consists of a large number of waves superimposed;

i.e., each forward stroke produces one waveform. When all the waves or traces coincide, the illusion of a stationary trace is obtained.

This can be arranged by synchronising the timebase with the voltage under examination, so that after a complete cycle the spot flies back and recommences to trace the second and succeeding cycles immediately on top of the first. It is usually impossible to make the timebase repetition frequency sufficiently stable to hold a pattern stationary for any length of time, and it is therefore necessary to apply automatic synchronising. This is accomplished by introducing a small portion of the test voltage to the timebase, so that the flyback cycle of the timebase is, in fact, initiated by the signal voltage.

A terminal marked 'sync' is usually fitted on the front panel of an oscilloscope for this purpose, being connected externally to the test signal applied to the Y plate. Internally, the 'sync' terminal is connected through a potentiometer which is also brought out at the front panel and sometimes marked 'sync control'. This potentiometer serves to adjust the amplitude of the Y signal applied to the timebase. Too great a sync signal may distort the waveform of the applied signal, while too little will not give a sufficiently stable lock.

Some oscilloscopes are equipped with a 'sync switch', which allows synchronising on either positive or negative waveforms, and which selects 'external sync', or directs to the timebase some of the Y voltage without the need for an external connexion to the 'sync' terminal, as required. Most modern instruments also incorporate a 'sync selector' control, allowing the timebase to be triggered either repetitively, as already described, or from a random occurring waveform. This latter condition is called 'triggering', where the synchronisation is automatic, as the horizontal trace is triggered by the Y signal. There are various associated controls which allow the sensitivity of triggering to be adjusted to suit the applied Y signal and waveform. Other features sometimes include an inbuilt integrator for synchronising the timebase to the field frequency of a television signal.

The timebase itself has facilities for varying the repetition frequency of the sawtooth waveform, and thus the horizontal speed of the spot across the screen. There are numerous types of timebase circuits, a popular version being designed around a Miller multivibrator circuit with d.c.-coupled paraphase output to the X plates of the tube.

The repetition frequency of the timebase can usually be varied from at least 10 Hz to 50 kHz, and most instruments have two controls, one giving switched frequencies over this range and the other being more or less a fine adjustment which provides complete coverage between the switched ranges. These controls are sometimes called 'sweep controls', whose calibration is in terms of microseconds or milliseconds per centimetre of screen swept. A typical sweep range is from 500 milliseconds per centimetre to 1 microsecond per centimetre.

TRACE EXPANSION

A useful feature is 'trace expansion'. This is a method of expanding the horizontal trace for detailed study of any part of the displayed waveform. For example, a normal line of the television Test Card C can be produced on the screen, and by expansion, can be lengthened horizontally to display the pulses which make up the 1·5 MHz frequency bars.

The expansion control regulates the gain control of an amplifier connected between the timebase and the X plates of the tube. Typical expansion varies the effective sweep length from 6 to 60 centimetres. The control is continuously variable from the centre of the screen up to 10 times the screen diameter. A maximum effective sweep of 60 centimetres, which the expansion control provides, could only otherwise be accommodated on a tube approaching 24 inches in diameter. Trace expansion also increases the effective sweep speed above the maximum provided by the timebase itself.

Y AMPLIFIER

If relatively small amplitude voltages are to be displayed usefully, an amplifier is required between the applied Y or test signal and the Y plates of the tube. This is called the 'Y amplifier'.

The inherent Y and X sensitivity of a cathode-ray tube is governed by several factors, namely: distance, d, between associated deflecting plates; length, l, of plates in the direction of the electron beam; distance, f, between the plates and the fluorescent screen; value of e.h.t. voltage, V_a, applied to tube final anode.

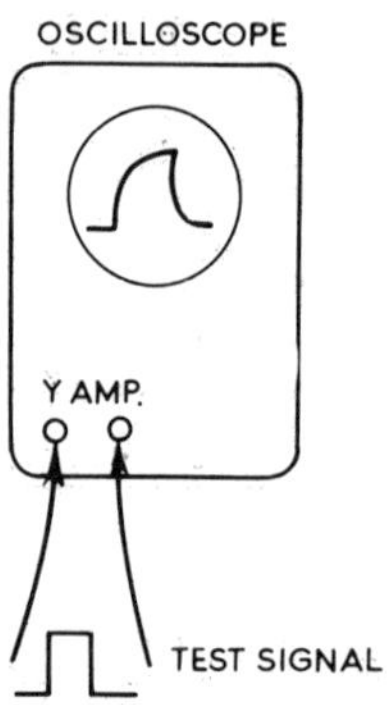

FIG. 6.5. *An insufficiently small rise time in the Y amplifier, due to limited high-frequency response, will distort a square wave as shown. The same effect will occur due to excessive capacitance between the connecting leads.*

Deflection is proportional to the potential applied across the deflecting plates such as:

$$\text{Deflection} = V_d . l . f / V_a . d$$

where V_d is the deflection potential, and the other factors are as given above.

The Y plates of a tube will not appreciably load the test circuit to which they are connected. In fact, connexion is possible to very-high-frequency circuits without undue trouble, provided that precautions are taken with regard to the inductance and self-capacitance of the connecting leads themselves. This also applies when the test signal is connected to the Y plates through a Y amplifier.

Moreover, the Y amplifier has to amplify the small test signals without distorting them, which, on square and pulse waveforms, means that the amplifier must have a relatively wide bandwidth. This is because square waves and similar transient pulses are made up of many harmonics of their fundamental frequency —it being the harmonic components which give waveforms their characteristic shape.

If such signals are applied to the Y plates through an amplifier which is unable to pass the higher harmonic components, then the display on the screen will not be a true indication of the test signal. There will be a marked rounding-off of the sharp corners of the waveform, as shown in Fig. 6.5. The same effect will occur if there is excessive capacitance between the connecting leads, even though the Y amplifier has a good high-frequency response, and particularly so if the test signal is derived from a high-impedance circuit.

Rise Time

Y amplifiers have a 'rise time' specification, which is the measurement of time that it takes the amplifier to respond to a sudden change in the amplitude of an input signal. If it were possible to cause a change from, say, zero to a maximum value of signal in zero time, the amplifier would require a finite time to respond to this change, and this is called the rise time of the amplifier (between the 10 and 90 per cent points).

The rise time is approximately related to the bandwidth of the amplifier in the following manner: $f_{3\,dB} = 0{\cdot}35/T$network, where $f_{3\,dB}$ is the upper frequency limit (MHz) at which the response is 3 dB below the mid-frequency response,

FIG. 6,6. *Excessive overshoot in the Y amplifier could be caused by sudden changes in response.*

and T network is the network rise time (μS). This reveals that a greater bandwidth is required for a smaller rise time specification. If, for example, it is required to display a square wave with a rise time of 0·25 microsecond, the amplifier will need a 3 dB bandwidth of no less than 0·35/0·25, which is equal to 1·4 MHz. Conversely, a Y amplifier with a 3 dB bandwidth of 3 MHz should have a rise time of the order of 0·35/3 microsecond, which works out to a little under 0·117 microsecond.

It is usual for a Y amplifier to have a high frequency response which is progressively diminishing. For example, the response may be 3 dB down at 3 MHz and 6 dB down at, say, 7 MHz. This gives a slightly smaller rise time than an amplifier which cuts off fairly sharply at 3 MHz.

To avoid 'overshoot' (see Fig. 6.6), the Y amplifier must have an even overall response which is free from any sudden change in amplitude.

VOLTAGE MEASUREMENT

The Y amplifier has an associated input attenuator which is usuall calibrated in volts per centimetre deflection. To determine the amount of deflection a graticule ruled in centimetre squares is fitted over the tube screen. Thus, if the attenuator switch is set to, say, the 1-volt-per-centimetre position, and an input signal gives a 2-centimetre vertical deflection, the signal amplitude is 2 volt peak-to-peak. This is 1 volt peak or 0·707 volt r.m.s.

Alternatively, a calibrated Y shift control may be used in conjunction with the calibrated attenuator. In this case, the voltage measurement is made by first setting the attenuator switch to a convenient value, so as to give approximately half-screen deflection (from top to bottom of the waveform in the case of an a.c. voltage). The calibrated Y-shift control is next set to zero and then the associated (non-calibrated) shift control adjusted until the top or the bottom of the waveform coincides with the horizontal centre line of the graticule. The calibrated Y-shift control is then adjusted until the opposite side of the waveform coincides with the horizontal graticule line originally selected. The peak-to-peak voltage is finally read from the appropriate scale on the calibrated Y-shift control.

To calculate the peak-to-peak value of an r.m.s. voltage applied to the Y terminals, the voltage must be multiplied by two times root two. For example, if the voltage is 6·3 V r.m.s., as may be obtained from the heater line of a receiver, the peak-to-peak value, *as indicated by the oscilloscope*, is $6{\cdot}3 \times 2 \times \sqrt{2}$, which works out to 17·8 volts.

Some oscilloscopes feature sockets on the front panel delivering an a.c. voltage of a known value which may be used for calibration purposes. More expensive models have a built-in square-wave oscillator which can be switched in for calibration checks. There are various other controls and terminals to be found on the oscilloscope which are adequately dealt with in the instruction manual supplied with the instrument. These include 'brilliance', 'focus', 'X shift', 'Y shift', 'astigmatism', etc., and their use is easily understood.

On most instruments a terminal is provided for the injection of a modulation voltage to the grid of the tube. If a sine wave is used for modulation, small blank intervals are produced on the trace which are useful for time markers and calibration purposes. Some of the more elaborate instruments feature a built-in oscillator which may be switched on as required to provide a similar facility.

On twin-beam and split-beam instruments the timebase and associated circuits are common to both beams, but there are usually two Y amplifiers, designated Y1 and Y2. The amplifiers may have identical specifications or the

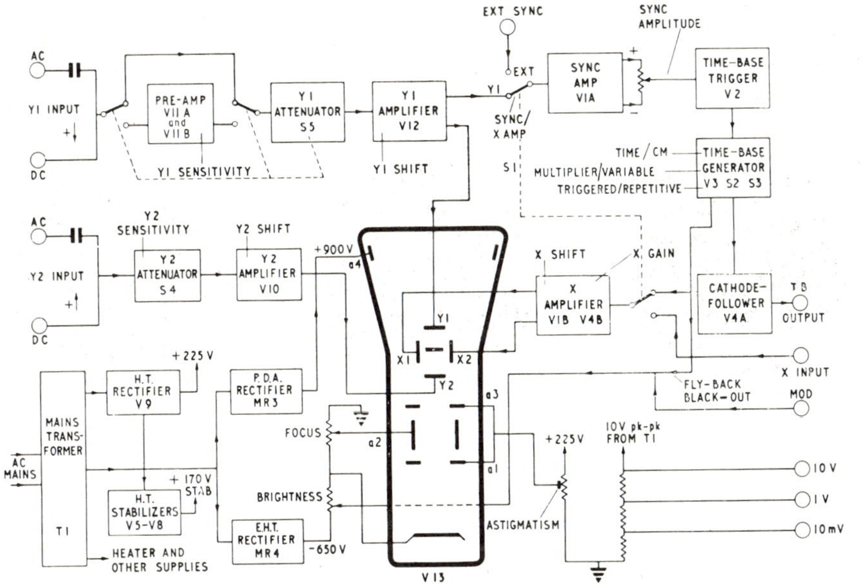

FIG. 6.7. *Block diagram showing the various features incorporated in the Cossor Model 1091 double-beam oscilloscope.*

Y1 amplifier may have a better bandwidth/gain factor than the Y2 amplifier, depending on the type of instrument. A block diagram of the Cossor Model 1091 oscilloscope is given in Fig. 6.7, which illustrates the salient features already discussed.

TIME AND FREQUENCY MEASUREMENT

There are several ways of measuring the frequency of a signal on an oscilloscope. A direct comparison of two frequencies may be obtained by feeding one to the Y input and the other to the X input. When the two frequencies are equal, an ellipse or circle will be displayed. If the frequency of the signal to the Y input is exactly twice that to the X input, two loops side-by-side will occur, as shown in Fig. 6.8*a*. For any number of loops, the Y frequency will be a multiple of the X frequency, given by the number of loops. For example, in Fig. 6.8*b* the Y frequency is half the X frequency, and in Fig. 6.8*c* the ratio is 2-to-3, that is, if the X frequency is 100 kHz, the Y frequency is 150 kHz.

If the two frequencies are not simple multiples or fractions of each other the pattern will revolve on the screen, with one series of waveforms rotating to the right and the other series rotating to the left. The speed of rotation of any one peak (that is, from any position back to that same position) is a measure of the frequency difference between the two signals.

Such patterns are called Lissajous figures, and can be used for frequency comparison, as described, up to ratios of about 10-to-1. On greater ratios it becomes difficult to count the number of loops, particularly if the pattern is rotating. The 50 Hz power supply frequency can sometimes be used as the known frequency, provided the unknown frequency is likely to be a multiple or simple fraction of 50 Hz. Supposing it is required to check the 300 Hz

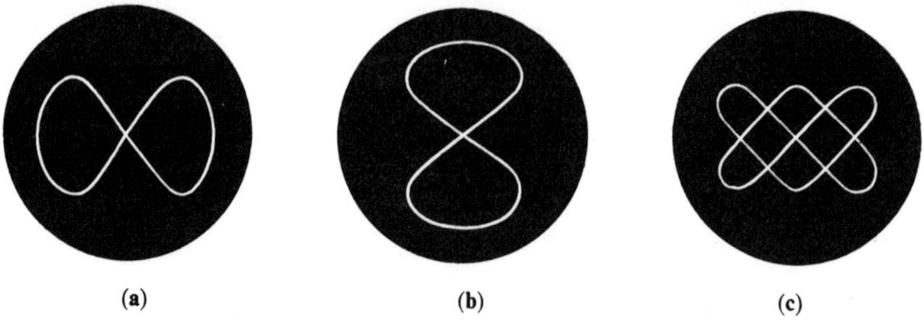

FIG. 6.8. *Lissajous figures obtained when the Y frequency is twice the X frequency* (a); *when the X frequency is twice the Y frequency, as at* (b); *and the figure at* (c) *indicating a frequency difference of 2-to-3: for example, when the X frequency is 100 kHz and the Y frequency is 150 kHz.*

signal produced by an audio oscillator; the power frequency should be applied to the X plates, via a suitable attenuator and transformer, and the oscillator signal should be fed to the Y plates, also attenuated or amplified as the case may be. The set-up is shown in Fig. 6.9*a* and the expected 6-to-1 Lissajous figure in Fig. 6.9*b*.

When the two frequencies differ considerably from a multiple or simple fraction of each other, the pattern is confused, and practice is required to recognise the sort of pattern being obtained while it is still revolving. When the pattern is stationary, though, the frequencies are an exact multiple within a few cycles per minute, thereby providing a basis for very accurate comparison.

In conjunction with an audio oscillator or generator, this method of fre-

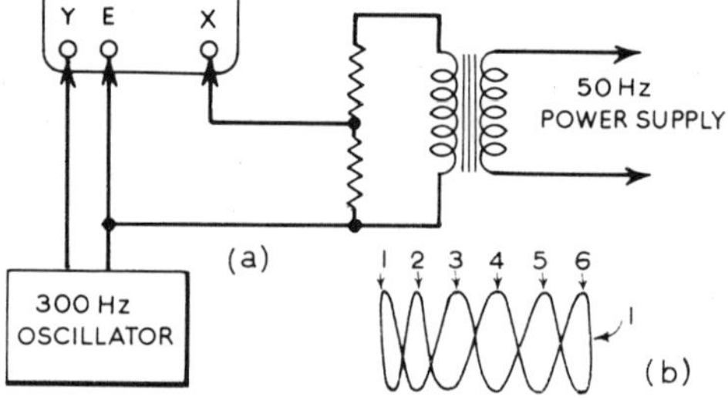

FIG. 6.9. (a) *The set-up which may be used to check the 300 Hz frequency of an audio oscillator by using the 50 Hz power frequency as a standard.* (b) *The 6:1 ratio Lissajous figure which should be obtained when the oscillator is working at 300 Hz.*

quency comparison may be advantageously adopted to adjust the frequency of an e.h.t. generator, tape recorder oscillator or similar circuit.

Another method of securing an approximation of the frequency of a displayed waveform is to measure its duration on the calibrated graticule. The X-shift control should be adjusted so that the start of the waveform coincides with a convenient centimetre mark, and the waveform (one complete cycle) should be measured along the X axis. The actual time measurement is obtained by multiplying the length of the waveform (in centimetres) by the time-per-centimetre as set on the sweep control.

To avoid error due to non-linearity at the extremes of the sweep, the sweep control should be adjusted to give several complete waveforms, and one in the centre of the trace should be selected for measurement. The frequency may be calculated by using the equations

$$\text{Frequency (Hz)} = 1/\text{time in seconds}$$
$$\text{Frequency (kHz)} = 1/\text{time in milliseconds}$$

An alternative method is to compare the two signals by applying one to the Y1 input and the other to the Y2 input when the timebase is operating.

PHASE MEASUREMENT

It is often necessary to measure the phase shift in an audio or r.f. amplifier or in an *LC* filter or similar network, and this may easily be accomplished with the oscilloscope. The idea is to apply the signal which is fed to the input of the amplifier to the Y input and the signal from the output of the amplifier to the X input, as shown in Fig. 6.10*a*.

The signal applied to the amplifier should be sufficient to drive it fully but not overload it, and under this condition the Y gain control should be adjusted to give about two-thirds vertical deflection. The Y input should then be disconnected and the potentiometer, in conjunction with the amplifier gain control, should be adjusted to give a like horizontal deflection. With certain oscilloscopes, the potentiometer can be dispensed with as the X gain control may be used to adjust the horizontal deflection (the timebase being switched off).

When the Y input is reconnected the resulting ellipse can be used to compare the phase relationship of the two signals, as shown in Fig. 6.10*b*. If the ellipse is positioned centrally about the graticule lines, the phase angle may be calculated from the expression

$$\sin \theta = A/B$$

Patterns over various phase angles are illustrated in Fig. 6.10*c*, and in the case of an amplifier the frequency may be varied over the passband to reveal how the phase changes with frequency.

This test should prove extremely interesting and useful to audio enthusiasts for checking the phase shift of their hi-fi amplifiers and for ensuring that the phase splitter is, in fact, feeding signals 180 deg. out of phase to each output

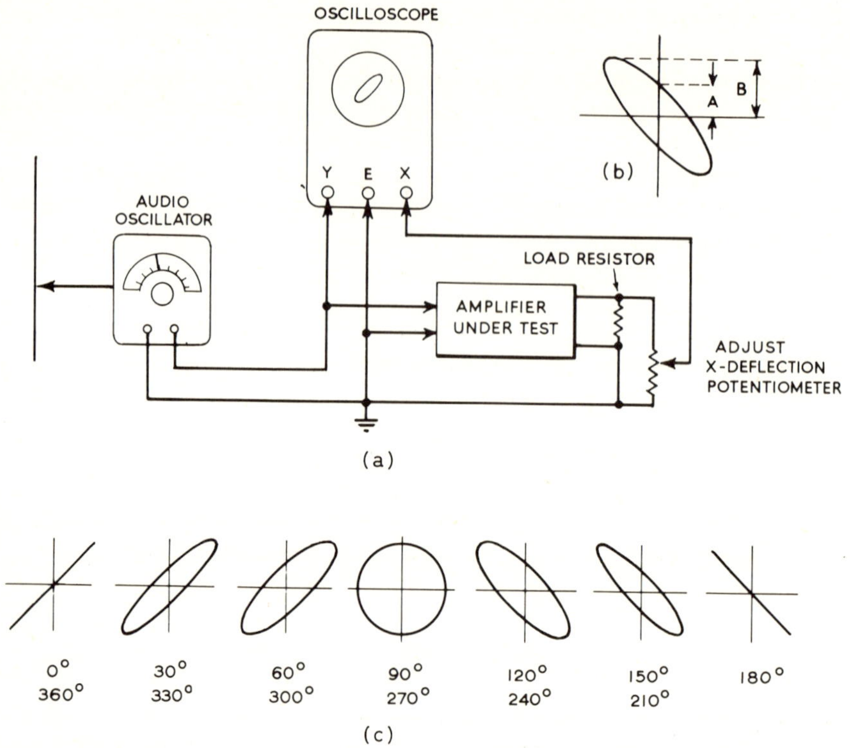

Fig. 6.10. (a) *Set-up for phase measurement.* (b) *Showing the method of determining the phase angle from the resulting ellipse. Patterns produced by various phase angles are shown at* (c).

valve. It is also useful for checking negative-feedback loops and the phase shift occurring in the output transformer over the audio spectrum. The phase shift occurring in filters, stereo decoders and like equipment may be checked in a similar manner.

It should be remembered that the oscilloscope amplifiers may themselves introduce phase shift if they are operated towards the limit of their passbands, and this should be taken into account.

VISUAL ALIGNMENT

Possibly one of the most useful attributes of the oscilloscope from the service technician's and experimenter's points of view is its ability to display the overall response curve of a radio or television receiver on its screen. This enables the state of tuned-circuit alignment to be *seen*, and in addition it reveals how the response is affected when adjustments are made to the various cores and trimmers.

To accomplish this a wobbulator is required, and the oscilloscope is set up so that the horizontal deflection is not a measure of time but a representation of changes in frequency. The vertical deflection is arranged to correspond to the output voltage from the receiver's detector stage. In other words, instead of a meter being used to indicate output (as in spot-frequency alignment), the vertical deflection of the trace is used.

If it be supposed that an ordinary signal generator is connected between the aerial and earth sockets of a receiver and that the generator and receiver are tuned to the same frequency, a vertical deflection will be obtained on the oscilloscope screen if the Y input is connected across the detector load and the generator signal is modulated. This deflection will alter in amplitude as the generator is tuned on either side of the frequency to which the set is tuned. This is of little use, however, as an ordinary output meter would give the same effect.

The next move is to deflect the trace horizontally, while at the same time changing the generator frequency plus and minus over the frequency to which the set is tuned, in step with each other. This can be done without difficulty with the wobbulator. We have already seen that this can be set to a fixed frequency and that a specific deviation either side of the fixed frequency is accomplished by means of an internal frequency modulator.

The deviation frequency—or frequency modulation—is often produced by feeding back to the wobbulator a portion of the sawtooth signal which is used by the oscilloscope for horizontal deflection. Thus, as the spot on the screen traverses from left to right the frequency of the signal applied to the set changes over a range as set by the wobbulator's deviation or sweep control. The idea is shown in Fig. 6.11, and a moment's reflection will make it clear that under these conditions a response curve (frequency versus output) will be traced to the screen. Fig. 6.12 shows the various components concerned in the formation of the response trace.

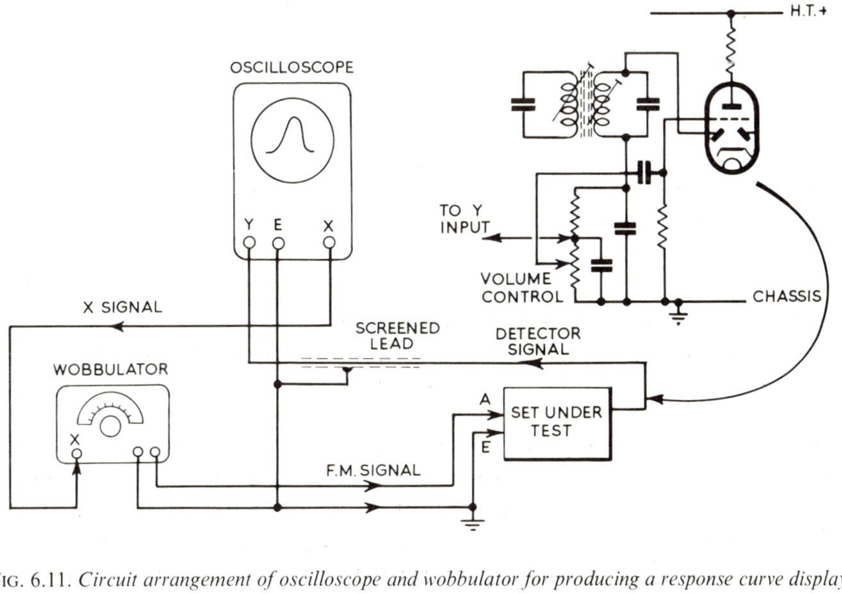

FIG. 6.11. *Circuit arrangement of oscilloscope and wobbulator for producing a response curve display.*

Connexion to the detector load in an ordinary radio receiver is usually to the top end of the volume control, as shown in Fig. 6.11. It will be understood that the deviation must be slightly greater than the passband characteristic of the tuned circuits and amplifiers of the set to produce a complete response curve on the screen. If the deviation is less, then only a part of the curve will be displayed, and if very much greater the curve will appear very narrow and give the impression of remarkable selectivity. The deviation can be adjusted at the wobbulator, as already described, and if the main frequency is set at, say, 1,000 kHz and the deviation for 20 kHz, the frequency applied to the set would vary linearly from 980 kHz to 1,020 kHz over the horizontal sweep.

For the alignment of ordinary a.m. receivers, a sweep of between 20 and 50 kHz is sufficient, which applies to the sound channel of television sets.

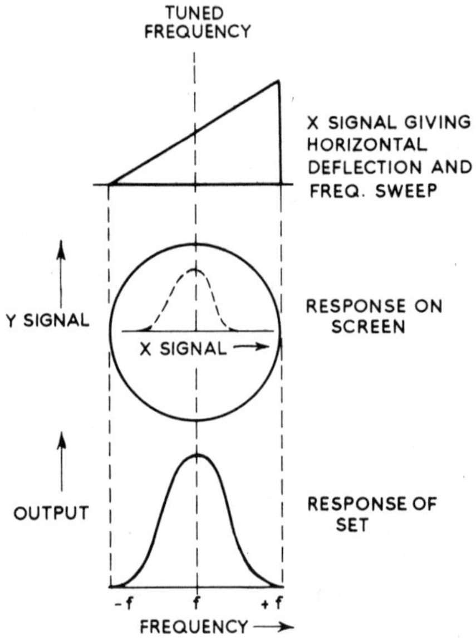

FIG. 6.12. *Various factors concerned in response display.*

With f.m. sets the sweep should be of the order of 200 kHz, for television vision stages 3 to 5 MHz, and for the overall response of a television receiver in the region of 8 MHz.

Some wobbulators are frequency-modulated from the power-supply frequency, and when these are used with an oscilloscope the timebase should be switched off and the power supply frequency also used for horizontal deflection. There is usually an 'X signal' terminal on the front panel of such wobbulators to feed the power-supply frequency from the wobbulator to the oscilloscope X terminal via a phase-correcting network.

When the power-supply frequency is used for horizontal deflection and sweep, the signal is of near sine-wave form and consequently a perfectly linear sweep, as with a sawtooth, is not obtained. Moreover, a displayed curve is produced on both sides of the sine wave (i.e., on the forward and return traces), and unless either one half-cycle of the sweep waveform is suppressed or the two displays are altered in phase so that they coincide, two responses may occur slightly displaced from each other. This does not affect the accuracy of the test in any way, but it can be disconcerting, and for this reason a phasing network is usually included in the wobbulator (or externally) so that the two responses occur one on top of the other.

A simple phase-correcting network is shown in Fig. 6.13*a* and in- and out-of-phase responses in Figs. 6.13*b* and 6.13*c* respectively.

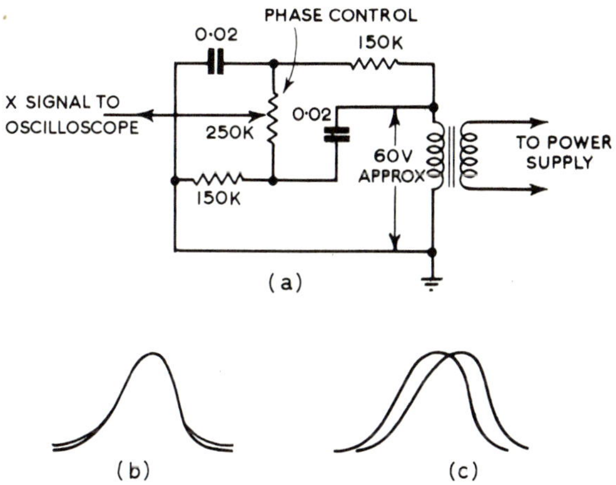

FIG. 6.13. *A simple phase-correcting network with two sets of response curves obtained.*

When the sawtooth waveform is used for horizontal deflection and sweep the timebase repetition frequency is not highly important, but if it is too fast, distortion of the response is likely to occur and if too slow flicker will be troublesome. It is found best in practice to use a frequency which is just high enough to avoid undue flicker of the trace.

MARKER SIGNAL

It is often desirable to identify certain points on the displayed response curve in terms of actual frequency. This is possible by injecting an unmodulated carrier signal from a separate signal generator to the receiver along with the sweep signal.

When the carrier signal is tuned to a frequency within the passband of the receiver under test a 'pip' will appear (resulting from a beat between the two signals) at the corresponding frequency point on the response display, and as the carrier frequency is altered so the pip will be seen to traverse the response curve.

If the amplitude of the carrier or marker signal is too great, the response display will decrease in amplitude and may be distorted. If the marker signal is too small it will be difficult to see the pip, especially if it falls on the sloping side

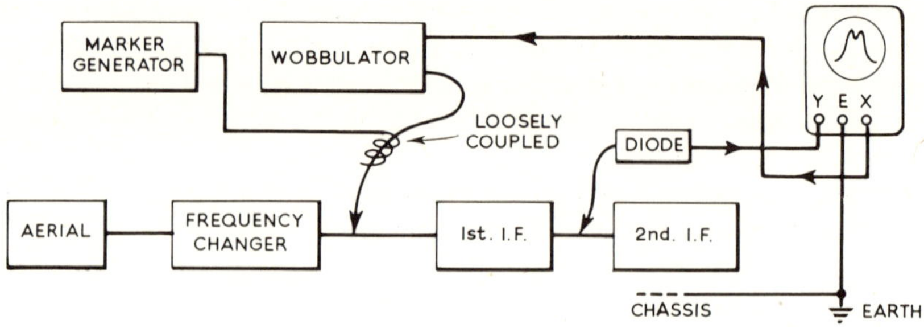

FIG. 6.14. *By the use of a diode probe the signal can be rectified at almost any point in the circuit for application to the Y amplifier. This illustration shows how the response of the first i.f. stage can be displayed by applying the signal to the input and the triode probe to the output.*

of the response curve. Experiment will indicate the best marker signal amplitude, and it is often possible to enhance the definition of the pip by connecting an 0·01-microfarad capacitor between the Y input and the earth terminal.

Some wobbulators feature an inbuilt marker generator, usually crystal-controlled, which gives a marker reference every 2 MHz or so depending on the fundamental frequency of the oscillator. This covers the whole band by the use of harmonics.

Other wobbulators carry a terminal on the front panel for the application of the marker signal from a separate generator or oscillator. If such a terminal is not fitted, it is a simple matter to inject sufficient marker signal to the point where the signal from the wobbulator is connected either via a resistor of about 120 ohms or through a small-value capacitor, the former method being preferred.

An alternative method is to coil a few turns of wire around the wobbulator signal input lead and connect the end of the wire to the marker generator r.f. output socket. This will give a very light coupling without affecting the loading

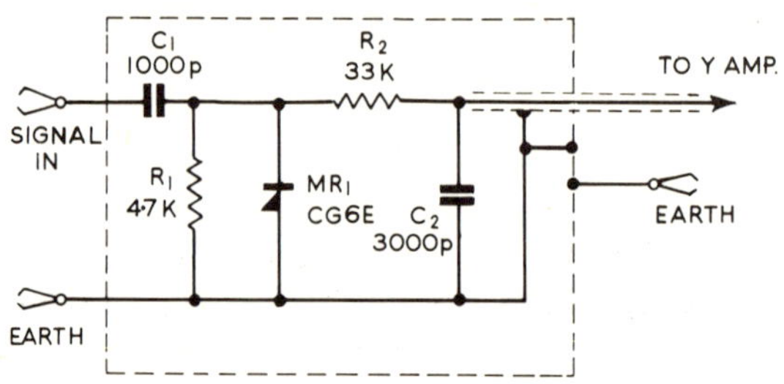

FIG. 6.15. *Circuit of a diode probe.*

to any marked degree, though it will usually be necessary to apply quite a strong generator signal to give a reasonable pip.

A simple, linear diode detector probe is sometimes supplied as an accessory with a sweep generator or wobbulator; or it may be integral to the instrument. Such a feature allows the r.f. or i.f. at almost any point in a receiver or amplifier to be rectified and applied direct to the Y amplifier terminal of an oscilloscope so as to provide a response display of the circuit between the point where the wobbulator signal is applied and the point where the diode is connected. This is made clear in Fig. 6.14. In Fig. 6.15 is shown the circuit diagram of a typical diode probe.

CHAPTER 7

Some Applications of the Oscilloscope

WITH THE ADVENT of colour television, the oscilloscope has become an essential instrument for the service workshop. Its value in television receiver servicing will possibly be appreciated, however, but not every user realises that this instrument can give very valuable assistance in dealing with radio and hi-fi audio equipment. In this chapter some of its more important applications are described.

RADIO RECEIVER ALIGNMENT

The usual procedure to be adopted is as already described for spot-frequency alignment. The i.f. stages are first aligned, and a marker signal is injected to provide a pip on the display at the centre i.f. The i.f. transformer trimmers or slugs are adjusted to produce a symmetrical curve like that shown in Fig. 7.1*a*, the idea being to achieve a response with sharply rising sides and a relatively flat top over the required passband.

With ordinary a.m. receivers or the a.m. section of a.m./f.m. receivers, the top of the response curve will automatically be comparatively narrow, depending on the coupling characteristics of the i.f. transformers. This is necessary in order to minimise adjacent-channel breakthrough; and in any case, an excessively wide passband is not required on a.m. sets. High-quality f.m. tuners and receivers would have a h.f. response approaching 15 kHz. With ordinary domestic-quality sets a smaller passband may well be desirable to provide sufficient selectivity and discrimination between the wanted and unwanted signals.

The best approach is to adjust all the trimmers to give a very 'peaky' response and then, if necessary, readjust the trimmers slightly to give the best symmetry and the required passband. If the passband is made too narrow, the audio reproduction will suffer owing to attenuation of the higher-order sidebands, but if too wide, medium-wave listening may be impossible owing to breakthrough of unwanted stations.

Once the centre frequency has been established on the marker generator, the frequency of this instrument should not be altered during the course of alignment. Various response patterns will be obtained as the trimmers are adjusted, and these will be relative to the marker pip; some of the patterns are

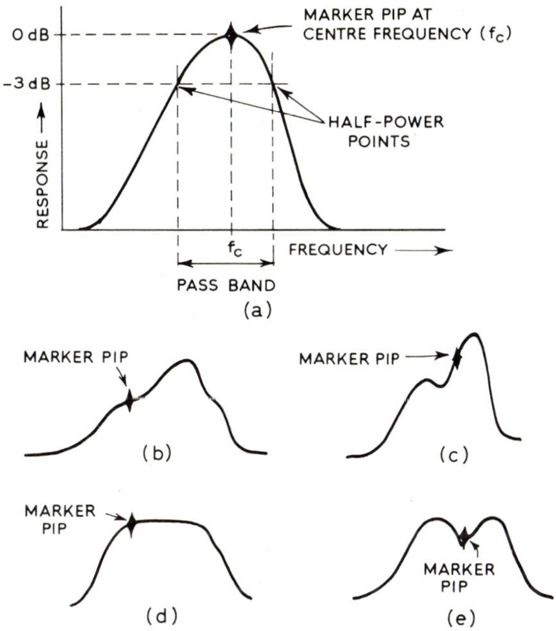

FIG. 7.1. *Displayed response curves with marker pip. At* (a) *the curve is symmetrical about the marker pip, which is set at the centre i.f. The half-power or 3 dB points are also shown here. At* (b), (c), (d) *and* (e) *are various displays which may occur during alignment.*

illustrated in Figs. 7.1*b*, *d* and *e*, but always the aim should be to secure a reasonably symmetrical curve as at Fig. 7.1*a*.

It pays from time to time to switch off the marker generator in case this happens to be affecting the response curve, which could happen if the marker signal is too strong. The marker generator r.f. output should be kept as small as possible, consistent with a marker pip of usable dimensions.

It is also most important to avoid overloading the receiver stages or the Y amplifier of the oscilloscope by too great a wobbulator signal or too much Y amplifier gain. Overloading from either of these causes will tend to distort the response display and probably give the trace the appearance of being beautifully symmetrical with a synthetically flat top, and thus lead the novice into trouble. After a good response has been obtained under one set of input signal conditions, the response curve should be viewed again with a smaller wobbulator input signal and a greater Y amplifier gain.

In some cases, the set's a.g.c. may be found to affect the curve, in which case it may prove advantageous to align with the a.g.c. disconnected, and finally to look at the response with the a.g.c. reconnected over various signal inputs. Some small trimmer correction may then be required for optimum results.

The local oscillator and r.f. stages can also be aligned visually by employing the marker pip for frequency calibration in relation to the tuning dial. However, for calibration purposes it is invariably best to adopt the normal spot-frequency techniques.

F.M. RECEIVERS

As alignment procedures recommended by manufacturers will vary from set to set, the following details should be regarded purely as a guide. The wobbulator should be connected to the input of the i.f. channel prior to the final i.f. or limiter stage, and should be tuned to the nominal i.f., as also should the marker generator (on most sets and tuners this is 10·7 MHz). The Y voltage may be picked up either from a diode probe connected to the final i.f. stage (Figs. 6.14 and 6.15), or from the a.f. take-off of the ratio detector or phase discriminator, but in the latter case the secondary of the detector transformer should be detuned so that the circuit may act as an ordinary a.m. detector or rectifier.

The secondary and primary windings of the associated transformer—that is the anode winding to which the signal is applied and the winding of the final i.f. amplifier or limiter—should be adjusted, in that order,to secure a response curve similar to that shown in Fig. 7.2*a*.

Alignment of the first i.f. transformer should be carried out with the wobbulator signal applied to the input of the mixer or frequency-changer. Where this point is not accessible, being inside the v.h.f. tuner, the signal may be capacitively coupled to the mixer either by winding a few turns of wire around the envelope and connecting this coil to the signal, or, in valve equipment, by lifting the screening can of the valve up the envelope away from its earthing and connecting the signal to the can. The secondary and primary windings of the associated transformer should then be adjusted to give a curve similar to that shown in Fig. 7.2*b*. It will be seen that this curve is narrower than that at *a* owing to the inclusion of two more tuned circuits, but when it has been obtained symmetrically about the marker pip, it should be flattened out by readjusting the core in the secondary winding until it resembles the oscillogram of Fig. 7.2*c*.

With the signal still applied to the mixer input, the core in the primary of the discriminator transformer should be adjusted to get a display like that of Fig. 7.2*d*. Finally, the core in the secondary winding, which may have been de-tuned initially, should be very carefully adjusted until a double curve as in Fig. 7.2*e* is produced. Optimum symmetry of the discriminator response occurs only when both diodes are in perfect balance in relation to their tuned circuits.

If a diode probe was employed for the production of a Y voltage, the probe should be removed for discriminator alignment and the Y voltage obtained from the a.f. take-off of the detector, usually across the volume control for convenience.

The above adjustments are best made with the wobbulator set to a sweep of about ± 300 kHz, for although this is in excess of the required passband (approximately 200 kHz for f.m. receivers, and not more than 250 kHz – 6dB points when a stereo decoder is fitted) it does make the curve of easily workable display width.

A reasonable assessment of discriminator linearity can be made by reducing the sweep to about ± 100 kHz, when a linear curve or line as shown in Fig. 7.2*f* should be obtained.

During the whole course of alignment the wobbulator signal output should

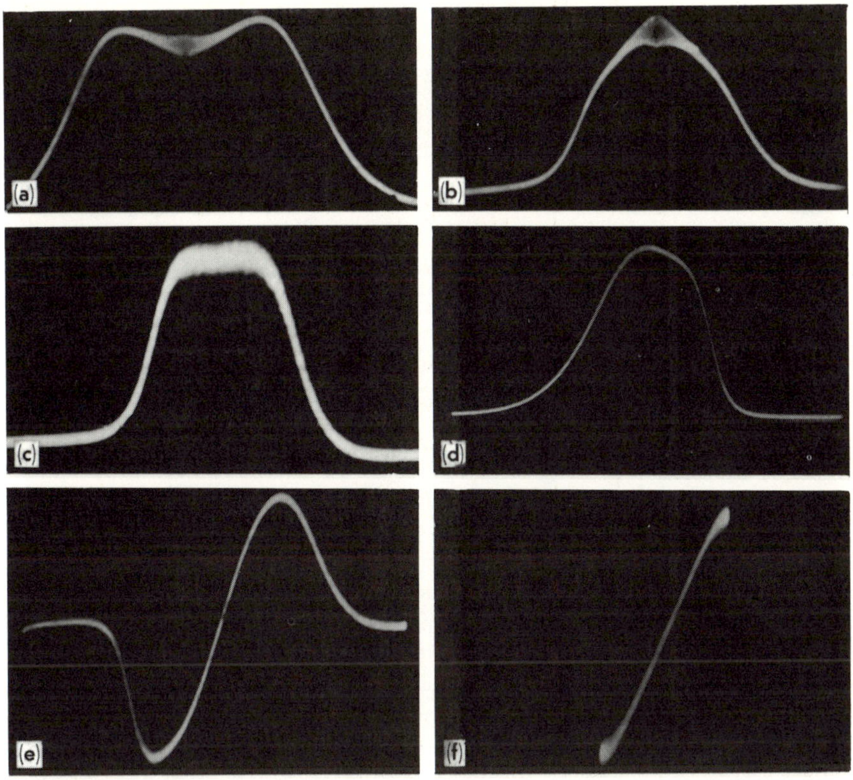

FIG. 7.2. *Oscillograms obtained during the visual alignment of an f.m. receiver. Marker pips are shown on curves* (a) *and* (b). *The marker generator should be adjusted to the nominal i.f., and the response curve should be resolved so that the marker falls at its centre frequency.*

be kept at the very minimum to avoid overloading. This may mean that a fairly substantial oscilloscope Y amplifier gain may be required to secure a response display of reasonable amplitude. It may also mean that slight 'noise' will be seen at the top of the curve, as for instance in Fig. 7.2*c*. This is of no consequence and the display is possibly more truthful under this condition than if a stronger wobbulator signal were applied.

There is one drawback, however, and that is high Y amplifier gain usually means that hum pickup will be troublesome. This can cause the trace to wobble up and down at a speed determined by the timebase repetition frequency. This can often be synchronised by adjusting the timebase to the hum frequency (50 or 60 Hz). It can also be greatly minimised by using screened cable to connect the Y signal to the Y amplifier of the oscilloscope. Such a precaution should always be adopted when the frequency of the Y signal is low and when capacitive loading is not likely to cause errors.

With combined a.m./f.m. receivers, the two alignment processes are totally independent of each other and, unless the makers' instructions detail the order in which they should be aligned, it matters little which section first receives attention.

Alignment of the r.f. and oscillator stages is best carried out by the spot-frequency method, particularly from the calibration aspect, but the overall response curve from the aerial terminals to the detector can be viewed to advantage by sweeping up to ±400 kHz and readjusting the r.f. and aerial trimmers or cores to get the very best results. Spurious responses falling slightly outside the i.f. passband can also be observed and cleared by sweeping over a greater range.

TUNED AND COUPLED CIRCUITS

The shape of the response curve due to two tuned and coupled circuits, as displayed on the screen of an oscilloscope, depends on the degree of coupling between the two circuits—known as the coupling coefficient or coupling factor. If the two circuits are lightly coupled and tuned to the same frequency, the response curve will be a single peak, such as that shown in Fig. 7.3*a*.

When the coupling is increased, as would occur by moving the windings closer together or increasing the value of the coupling capacitor, the voltage induced across the secondary winding increases until a point is reached where the voltage remains substantially constant. Further increase in coupling causes the top of the response curve to flatten out, as shown in Fig. 7.3*b*. Still further increase in coupling results in the response curve breaking into two peaks with a trough between, as shown in Fig. 7.3*c*. These are the three essential

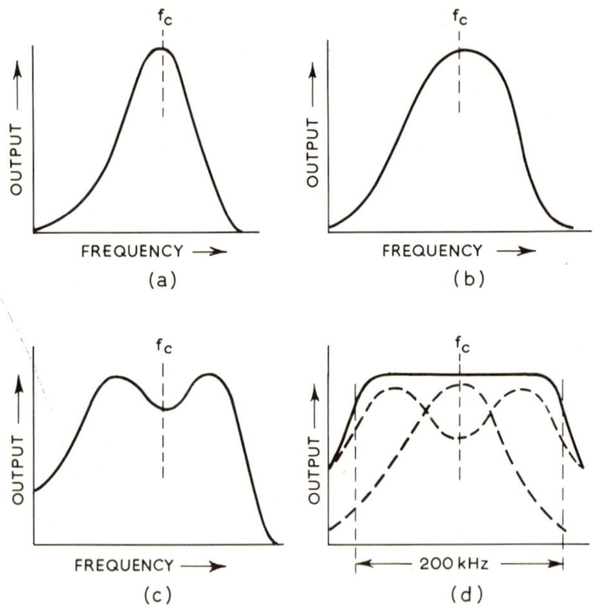

FIG. 7.3. *Response curves obtained due to transformers:* (a) *below critical coupling;* (b) *at critical coupling and* (c) *above critical coupling. At* (d) *is shown how the responses of two pairs of tuned circuits (below and above critical coupling) are added to produce an overall response curve with sharply rising sides and a relatively flat top.*

modes of coupling used in radio and television receivers (in i.f. transformers, for example).

The coupling coefficient at *b* is known as the 'critical' value, at *a* below critical and at *c* above critical. All three modes will at some time be displayed on the screen of an oscilloscope, so it is as well to understand them. It should be noted, however, that if the primary and secondary windings of, say, an i.f. transformer are not tuned to the same frequency a response similar to that of an over-critically coupled transformer may result. This method is, in fact, sometimes adopted to increase the bandwidth or passband of an i.f. stage. Another method is to 'stagger-tune' the various circuits so that the combined response curve has sharply rising sides and a relatively flat top. In some f.m. receivers two responses due to a greater-than-critical and a less-than-critical coupling are added to produce the overall response curve, as shown in Fig. 7.3*d*.

TELEVISION RECEIVERS

The sound channel of 405-line television receivers may be visually aligned in the same manner as ordinary broadcast receivers. The i.f. wobbulator signal and marker generator signal must, of course, equal the sound-channel i.f., and the signals should be injected at the mixer or frequency-changer.

To avoid undue breakthrough of the vision signal on sound, it is best to peak the sound, i.f. transformers and leave the general coupling coefficients of the tuned circuits to provide the sound-channel bandwidth. A special attempt to broaden the sound passband, having in mind possible improvement of the sound quality, does not usually pay off, since then the sound response is so wide that it allows some of the vision sideband components to get into the sound channel, thus giving rise to the characteristic vision buzz. However, a too peaky sound i.f. may cause instability and impair the sound interference limiting action, thereby requiring a gain/bandwidth compromise on certain sets.

The intercarrier sound channel of 625-line standard receivers must have a response characteristic whose centre frequency falls exactly at 6 MHz, with symmetry either side. This channel is always best aligned visually, for like any f.m. receiver, distortion can result from asymmetry of the characteristic either within the intercarrier channel or at the f.m. detector. For the alignment of this section, therefore, reference should be made to the section dealing with the alignment of f.m. receivers.

The Vision Stages

405-line vision stages require a sweep of some 4 to 5 MHz. The Y amplifier terminal of the oscilloscope should be connected to the 'live' side of the vision detector load resistor and the 'earth' terminal should be connected to the set's chassis—taking isolation precautions in the case of possible 'live' chassis equipment, as already detailed.

The wobbulator and coupled marker signal should be applied between the input of the frequency-changer and receiver chassis, and it is best to terminate the wobbulator by an 82-ohm resistor, except when connexion is made direct to the aerial socket of the receiver under alignment. The contrast (and gain control if fitted) should be set to maximum.

The wobbulator tuning should be adjusted so that it 'sweeps' the vision i.f. passband. Tuning the wobbulator will cause the response display to shift across the screen of the oscilloscope and adjustment should be made so that the display appears in the centre of the tube, taking care to ensure that the X shift control is properly centred to give a full trace across the screen. The sweep or deviation control on the wobbulator should be adjusted to give a curve which occupies approximately two-thirds (or more) of the screen diameter. A narrow curve means too much sweep and a wide curve, which cannot be fully contained on the screen, means too little sweep.

Difficulty may be experienced in obtaining a curve of any sort. In this event, a fairly strong wobbulator signal should be applied and the oscilloscope's Y gain should be advanced. Tuning the wobbulator over the set's vision i.f. should then produce a response of excessive amplitude, with the top flattened owing to overloading. Spurious responses at frequencies removed from the vision i.f. sometimes occur when the wobbulator is tuned over the band, and care should be taken to ensure that the response curve which has the largest amplitude (provided it is in the correct i.f. range) is selected.

The wobbulator output signal should then be decreased until the characteristic shape of the top of the curve appears (i.e., until the synthetic flatness disappears). This may also necessitate a reduction of Y amplifier gain, and it is always best to work with as little wobbulator signal as possible and to maintain the required vertical deflection of the trace by Y gain. This can only be taken to the point where trace wobble, due to hum pickup, cannot be eliminated, and becomes disturbing. Of course, if the i.f. stages are very much out of alignment, then a greater wobbulator signal will be needed, but the wobbulator output should be progressively decreased as the various tuned circuits are brought into alignment, in accordance with normal alignment practice.

At this stage it will pay to get used to the various response effects which can be obtained by altering the tuning cores of the vision i.f. transformers, after first identifying the coils from a service sheet or manual of the receiver under adjustment. The effect that the various controls of the oscilloscope and wobbulator have on the display should also be carefully observed, and this applies also to the marker generator and signal amplitude.

With the controls properly adjusted and the wobbulator carrier and sweep control set so that the applied signal is 'wobbulated' within the vision i.f. passband, a response curve similar to that shown in Fig. 7.4 should resolve if the i.f. channel is tuned correctly. This curve is based on actual v.h.f. carrier frequencies (Channels 4 and 2), but would apply equally at the vision and sound i.f's. The trough at the sound frequency is caused by the sound rejectors in the

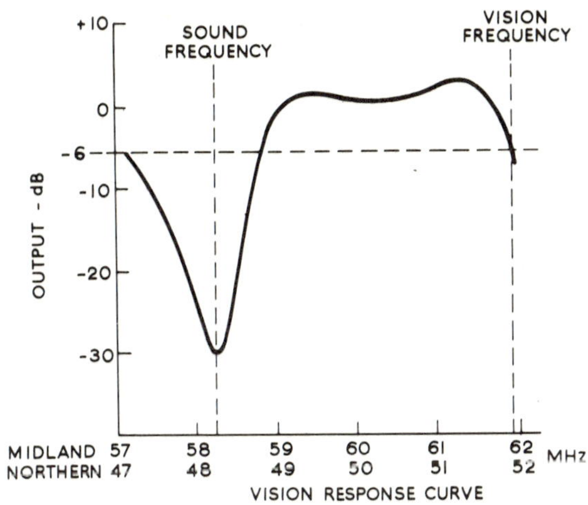

FIG. 7.4. *Vision response curve, showing the trough at the sound frequency caused by the sound rejectors. The vision carrier frequency falls on the sloping side of the vision response curve at a point 6 db down from maximum.*

vision channel, and it should also be noted that the vision carrier frequency falls on the sloping side of the vision response curve at a point 6 dB down from maximum, this being necessary to obtain the correct reception characteristics from a single-sideband vision signal.

The marker signal should be adjusted to the sound carrier frequency and the vision carrier frequency to ensure that the pip occurs at the correct points on the response curve. It is imperative that the frequency of the marker generator is very accurate for these tests, as the whole alignment process requires the response curve to be built up in relation to the marker pips.

If necessary, the marker generator should be accurately calibrated at the sound and vision carrier frequencies by the 'beat frequency' method illustrated in Fig. 1.10.

The overall bandwidth of the i.f. channel can be determined by adjusting the marker generator so that the marker pip traverses the curve between the half-power points (see Fig. 7.1). The frequency over which the marker generator has to be tuned to achieve this represents the passband in MHz.

It is not possible to provide detailed instructions for aligning a television receiver as each manufacturer favours his own methods to suit the particular circuit design. The object is, however, always to secure a vision response curve as close as possible to that in Fig. 7.4. The exact shape will vary slightly from one receiver to another, but the basic frequency and output characteristics should be within about 10 per cent of those illustrated.

With 405-line receivers the vision carrier i.f. is around 34·5 MHz and the sound i.f. at 38 MHz. This means that the local oscillator is set to work at a frequency which is equal to the incoming carrier frequency *plus* the i.f. (i.e. oscillator high). The oscillator of early models may be below the signal frequency (i.e. signal frequency *minus* the i.f.), which makes the vision i.f. at a higher frequency than the sound i.f. (opposite to the former example). This point should be remembered when viewing the displayed vision i.f. response curve, to ensure which side of the curve should fall to the base of the trough representing the sound carrier i.f. It is rather important, therefore, to check on the vision carrier i.f. in the service manual or service sheet before setting the marker generator.

Before attempting alteration to any of the tuning cores, a selection of insulated trimming tools should be available to suit the various sizes of slots, stars and hexagon shapes used in the core ends. The cores should also be eased in their formers to rotate freely. Various methods are used for sealing cores. Some manufacturers use a greasy substance, while others use a thin length of rubber or elastic inserted between the core and former. This latter method is favoured. After adjustment the cores should be re-sealed to prevent vibration from shifting the adjustment.

On no account should a trimming tool which does not fit snugly into the core be used, as the core will most certainly be damaged and possibly lock into the former. In the event of this happening (which it never should!), the tool may be removed either by drilling it out or by breaking it up until it is loose enough to fall out of the former. This is a tedious task and not always successful, since the former may break and require replacement of the coil.

Some alignment processes require the use of damping resistors and/or capacitors. These should always be of non-inductive type, and should be carefully terminated with crocodile clips as shown in Fig. 7.5. The damping components necessary will vary from receiver to receiver and it is best to produce these as and when required.

The tendency for instability to result owing to the connexion of the oscilloscope to the vision detector circuit can be reduced by connecting a 47k resistor in series with the Y amplifier input.

It is sometimes recommended that the vision a.g.c. circuit be made inoperative while alignment is being carried out. This can be accomplished by connecting a

FIG. 7.5. *The use of crocodile clips for load terminations.*

3-volt battery between the chassis and the a.g.c. line, with the positive terminal of the battery to chassis.

Fig. 7.6 shows a series of response curves taken during the alignment of a television receiver with a 34·65 MHz vision i.f. and a 38·15 MHz sound i.f. Trace *a* was obtained by applying the wobbulator signal to the control grid of the final vision i.f. valve, with a sweep over 33–38·5 MHz, and adjusting the cores of the i.f. transformer between the anode of that stage and the vision detector.

The marker signal is at 34·65 MHz which, when the i.f. transformer is correctly adjusted, should produce a pip on the display 6 dB down the curve, as shown. Adjustment was also made to get the top of the curve as flat as possible.

Trace *b* was obtained by applying the wobbulator signal to the control grid of the *first* vision i.f. valve and adjusting the cores of the inductors in the grid circuit of the *second* i.f. valve. At this stage the sound rejectors were also adjusted. There are two ways of doing this: one is to disconnect the wobbulator signal and in place inject a modulated a.m. signal at 38·15 MHz (sound i.f.) from the marker generator if suitable. This will produce a waveform on the screen of the

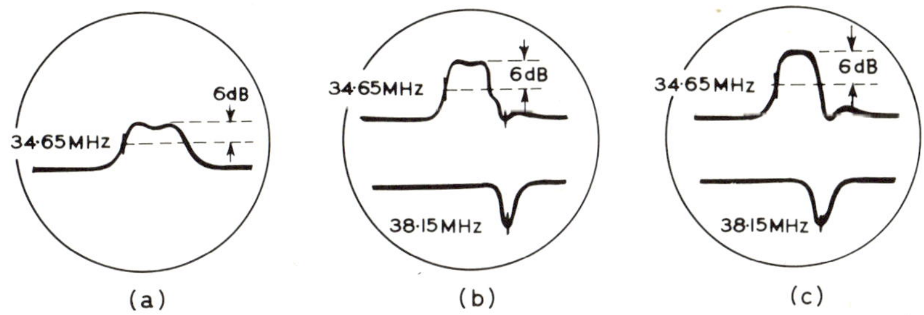

FIG. 7.6. *Series of response curves produced during the alignment of a television receiver. The traces at* (a), (b) *and* (c) *are described in the text.*

oscilloscope, the amplitude of which should be brought to minimum (increasing the input signal if necessary) by adjusting the cores in the sound rejector coils.

The second way is to maintain the wobbulator input and set the marker generator to the sound i.f. The pip will then fall at the base of the trough, at the sound carrier point, as shown at *b*. The maximum amplitude trace of the response should then be obtained, and the section in close proximity to the trough should be expanded. This can be achieved either by reducing the sweep and adjusting the wobbulator frequency very slightly so that the trough section falls in the centre of the screen, occupying almost the full screen width, or by X expansion, if provision for this is featured on the oscilloscope.

When the trough section is fully displayed the wobbulator input signal and/or Y amplifier gain can be increased to expand vertically that section of the curve. It may also be necessary to adjust the marker signal amplitude to get a good marker pip without trace distortion. The rejectors should then be very carefully adjusted to give the greatest depth of trough while keeping the marker pip right at the base. Trace *b* shows such a pip and also a pip corresponding to the vision carrier at the 6dB point on the curve. In order to obtain two pips two marker signals are required, but a single marker signal is usually sufficient in practice as the frequency of this can be adjusted as required. Also shown in *b* is the sound response curve in relation to the sound carrier pip.

Trace *c* was obtained with the wobbulator signal fed to the tuner test point, which is really at the signal grid of the frequency-changer or mixer valve in the tuner, a test point being made available on most tuners for ease of connecting the signal. The symmetrical curve of *c* can be obtained by adjusting the tuner i.f.

transformer core and the core or cores of the transformer or inductor in the control-grid circuit of the *first* vision i.f. valve (this may be the i.f. valve common to both sound and vision).

The adjacent-channel rejector, which is comparable to an ordinary sound rejector, should be adjusted in the same manner as the sound rejectors, but at the appropriate frequency. In the case of a set with i.f.s as given, the adjacent-channel rejector should be adjusted at 33·15 MHz, to which frequency the marker generator should also be tuned, to identify the response at the adjacent-channel point.

Adjacent-channel selectivity relies on this adjustment, and is most important in areas where adjacent-channel television stations can be received; also where the set is to be used on a 'coaxial relay system', on which a master aerial feeds a multiplicity of signals to subscribers, sometimes on adjacent channels. Adjacent-channel interference takes the form of vertical lines superimposed on a picture, having a 1·5 MHz spacing.

A filter tunable over the i.f. is usually situated in the aerial input lead to the tuner. This can be adjusted to provide maximum rejection to any strong interfering signal within its range of adjustment.

Finally, the overall response curve should be viewed by applying the wobbulator signal, at the appropriate channel frequency with a sweep of 3 to 6 MHz, to the aerial socket of the receiver. If the top of the curve has an excessive tilt, this may be corrected by adjusting the slug in the r.f. coil biscuit of the tuner. The tuning of this core is very flat and if the core is turned too far in one direction it will become disengaged from its thread in the former. Should this occur, the appropriate biscuit must be removed from the tuner in order that the core may be re-engaged. The final display should very closely approximate the curve of Fig. 7.4.

After the receiver has been connected to the aerial and switched on for several minutes, the correct sound and picture should be obtained on all channels with the fine tuning control close to the centre of its travel. However, if it is necessary on any channel to set the tuner well away from the centre of its travel, adjustment of the oscillator slug is required, with the fine tuner at centre, for maximum sound consistent with minimum sound-interference-on-picture and best picture quality.

Dual Standard, 625-Line and Chroma Alignment

Dual-standard receivers are somewhat more complicated to align than 405-line-only models, but the principles already expounded are still applicable, particularly to the 405-line section of the i.f. channel. However, the i.f. channel design is generally focused towards optimum 625-line response, with extra tuned filters and rejectors being switched in on the 405-line standard to reduce the passband, to add sound signal rejection and to tailor the response characteristic generally. The sound channel always operates as the 405-line sound i.f. on the v.h.f.

channels and as the intercarrier channel on the u.h.f. channels. Of course, there are switchable sound detectors for a.m. demodulation on the 405-line standard and f.m. demodulation on the 625-line standard.

625-line-only sets are no more complicated to align than their 405-line-only counterparts. The trend is for the filtering and selectivity to be built into the first one or two stages of the i.f. channel and for the gain to be provided by the subsequent stages. This technique is desirable in transistor receivers where intermodulation and crossmodulation can result from a multiplicity of signals. Thus the number of signals is reduced by early filtering.

The one i.f. channel accommodates the whole of the vision signal sidebands *and* the sound signal, but the latter is attenuated so that it arrives at the vision detector about 40 dB below the vision carrier. The 6 MHz beat between the sound and vision carriers yields the intercarrier signal and this is then passed into the intercarrier sound channel which is terminated with the f.m. detector. Other filters are incorporated in the i.f. channel to eliminate adjacent channel signals. It is thus necessary to consult the makers' service manuals for the correct alignment procedures, and most nowadays recommend visual alignment with a wobbulator and oscilloscope.

This is particularly the case with colour receivers where very accurate alignment is essential for the correct colouring. We have seen that on the 405-line standard the vision carrier is placed about 6 dB down on the falling side of the vision i.f. response characteristic to compensate for the vestigial sideband characteristics of the transmission. The same applies to the 625-line standard, for this is also of the vestigial sideband nature. However, with colour we have extra signal energy in the form of the chroma subchannel components, and in some receiver designs the subcarrier also has to fall about 6 dB down on the slope at the other side of the i.f. response characteristic. Excessive carrier here can precipitate quadrature distortion. Moreover, the phase characteristics of the i.f. channel are more critical in colour receivers than in simple monochrome models, and these are also determined by accurate alignment as well as by good design.

The bandpass chroma amplifier might also incorporate tuned circuits, and when separate detectors are employed for the luminance and chroma signals (not meaning the synchronous detectors of the PAL decoder) at the conclusion of the i.f. channel, it is not uncommon for the chroma bandpass amplifier to be aligned for vision carrier enhancement as a means of minimising quadrature distortion, with a 6 MHz rejector to notch out the intercarrier signal.

The luminance channel also features a chroma subcarrier rejector (tuned to 4·43 MHz) which has to be aligned accurately. The decoder, too, incorporates tuned circuits at the chroma subcarrier frequency and at half the frequency of the line timebase to produce the ripple signal for synchronising the PAL switch.

Clearly, it is impossible in this volume to explore all the tuned circuits and alignment procedures relevant to dual-standard, 625-line and colour receivers. Readers wishing to obtain more information on these subjects might care to refer to my companion books by the same publisher entitled *Television Servicing Handbook* and *Colour Television Servicing.*

WAVEFORMS

The first waveform that one ever sees on the screen of an oscilloscope is the waveform of the power supply at 50 Hz (60 Hz in the United States). There is always a residual mains pickup in proximity to an oscilloscope, and with a reasonable setting of Y amplifier gain this is sufficient to provide a vertical deflection, and hence a sine wave when the timebase is running. The sine wave is usually a little distorted from this source, and its amplitude on the screen can be made to increase significantly by holding a finger on the Y amplifier terminal or by holding a wire connected to that terminal, depending on the Y amplifier gain setting.

The sweep velocity or repetition frequency should be adjusted on the coarse and fine controls until there appears on the screen a stationary trace showing one cycle of the applied signal waveform, some whole number of cycles or some simple fraction of a cycle. When the timebase has the same frequency as the applied signal, just a single, almost complete cycle will be displayed. The end may be clipped slightly due to synchronisation. If excessive sync is applied, then only a part of the full cycle will be displayed. A small part of the waveform will never appear owing to the finite time of the re-trace or flyback.

If the sweep takes, say, three times as long to occur as the full cycle of the applied signal, then the spot will trace through three full cycles, and three cycles will appear on the screen. In fact, the number of cycles appearing on the screen can be found by dividing the signal frequency by the repetition frequency of the timebase.

If the timebase frequency is higher than the applied signal frequency, say twice as high, there will occur two horizontal sweeps during each signal cycle. On the first sweep will be traced the first half-cycle of the signal and on the second sweep the second half-cycle. These two sections will appear simultaneously on the screen owing to persistence of vision and the repetitive trace accuracy (Fig. 7.7*a*). The precise manner in which the waveform is split will depend on the phase relationship between the sweep and signal frequencies.

Because it is impossible to display one absolutely complete cycle on the screen, due to re-trace etc., it usually pays to set the timebase to half the frequency of the applied signal, thereby giving two cycles on the screen. In this way one complete cycle may then be observed from start to finish, and this is desirable when investigating a complex waveform or checking for distortion. See Fig. 7.7*b*.

The points at which a single cycle starts and finishes on the screen depend on the phase relationship between timebase and signal frequencies. Where the phase is correct, the waveform will appear as shown in Fig. 7.7*c*, and where there is a phase difference the display may be as in Fig. 7.7*d*. The phase relationship may be altered to produce the conventional sine-wave display—shown in Fig. 7.7*c*, which also reveals that part of the cycle is cut off owing to re-trace—by making a slight adjustment to the fine sweep control and/or the sync control.

It is interesting to note that as the frequency is increased, more will be cut off the single-cycle display. This is because the re-trace then becomes a relatively larger part of the cycle.

Also, the higher the repetition frequency the less bright will appear the trace on the screen. This is because of the high speed of spot travel on the screen, and it can be compensated for by increasing the setting of the brightness control. It is for the same reason that the re-trace or flyback (when flyback suppression is not used) always appears somewhat less intense on the screen. Normally, however, flyback suppression is used, and this operates by a negative pulse being fed back to the control grid of the cathode-ray tube during the flyback period. Some oscilloscopes also feature an automatic brightening control.

For some tests it may be desirable to observe the waveform during the flyback as well as during the forward trace. This can be accomplished by switching off

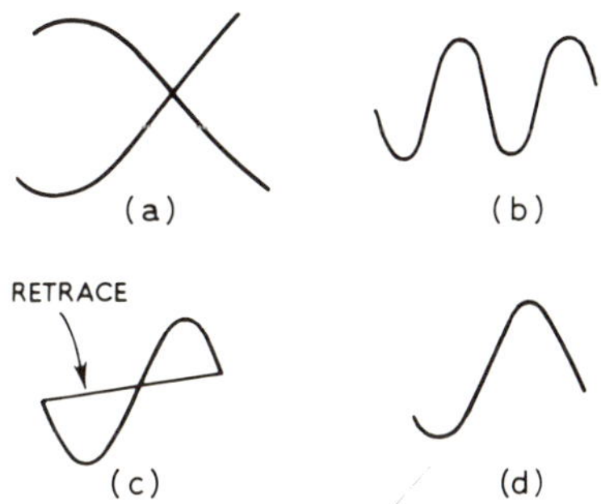

FIG. 7.7. *The pattern at* (a) *may occur when the timebase repetition frequency is twice the signal frequency. The two-cycle display at* (b) *allows the examination of one complete cycle. At* (c) *is shown a single-cycle display without phase shift (note the period occupied by the retrace and consequent suppression of part of the cycle). When there is phase shift between the repetition and signal frequencies, a single-cycle display may resolve as at* (d).

or disconnecting the flyback suppression and turning up the brightness control so that the re-trace is reasonably visible. Such a procedure may be adopted to observe one or two hidden cycles when a very high-frequency signal is applied to the Y amplifier and the timebase is set to a lower repetition frequency—it may, in fact, be running at full frequency under such a condition.

When a signal waveform is to be examined, the coarse and fine sweep controls are adjusted to display one or more waveforms, as already described. The fine control should then be carefully rotated until the display locks. At low frequencies this is not unduly difficult to achieve, but at high frequencies the fine control may appear extremely critical. However, when the sync control is advanced the timebase will come under the influence of the synchronising voltage, and the display will become stationary. On instruments with automatic triggering, and with the associated 'stability' control set correctly, the applied waveform will lock on one or any number of waveform displays as the sweep controls are adjusted.

On instruments with ordinary sync control, the setting of the control will regulate the amplitude of the signal applied to the timebase (this connexion may be made externally from the Y terminal to the 'sync' terminal or internally by operation of the 'sync' switch). Too little sync amplitude may not lock the display adequately, while too much sync amplitude will distort the display, and may even change the timebase frequency sufficiently to break the display into several juddering sections. The correct procedure is to advance the sync control to a very low setting and then adjust the sweep controls to hold the display almost stationary, after which the sync should be further advanced just to the point where the display is perfectly locked.

On instruments with auto-triggering, the trace will only occur when a signal is applied to the Y amplifier. There may be two continuously variable controls: one marked 'trigger repetitive' and the other marked 'trigger sensitivity'. The former progressively changes the timebase from the trigger to the repetitive mode, while the latter increases the sensitivity of the triggering mode.

Under normal conditions, the former control would be adjusted towards the 'trigger' side until the timebase just cuts off (possibly leaving the spot stationary on the extreme left of the screen). The 'trigger sensitivity' control should be turned right down and the signal applied to the Y amplifier terminal. Finally, the 'trigger sensitivity' control should be slowly advanced to a point where the timebase just starts. This is the most sensitive condition, and the timebase will then lock automatically to the input signal.

Modulated Waveforms

A typical modulation waveform can be displayed simply by connecting the Y terminal of the oscilloscope to the output of a modulated signal generator—ensuring that the 'earths' are also connected, of course. The Y amplifier of the oscilloscope will need to respond to the r.f. component of the modulated signal from the generator, and the generator frequency should be adjusted to a frequency within the Y amplifier range.

The display will lock on the modulation frequency and a pattern similar to that shown on the lower trace of Fig. 7.8*a* will resolve. This is known as the 'modulation envelope' and the waveform on the top trace is the a.f. signal which was used to modulate the r.f. carrier. It will be seen that the modulation waveform closely matches the contours of the modulation envelope, thus indicating very little modulation distortion (or envelope distortion). A double-beam oscilloscope was employed for these displays with the Y1 amplifier connected to the modulated r.f. source and the Y2 amplifier connected to the a.f. source. These displays were produced with a 2-MHz r.f. signal modulated by a 400-Hz a.f. signal during the course of development of a carrier-generator/modulator system for coaxial relay systems.

The waveforms in Fig. 7.8*b* show severe envelope distortion which was produced at the commencement of the development owing to the carrier-generator oscillator not being able to follow the positive peaks of the modulation signal.

MODULATION PERCENTAGE

Modulation depth, as well as envelope distortion, may also be discovered from such a display by first measuring the vertical distance between the envelope peaks and then measuring the distance between the troughs. This gives a maximum and minimum amplitude, and the minimum should next be subtracted from the maximum, giving an amplitude difference. The amplitude sum is then

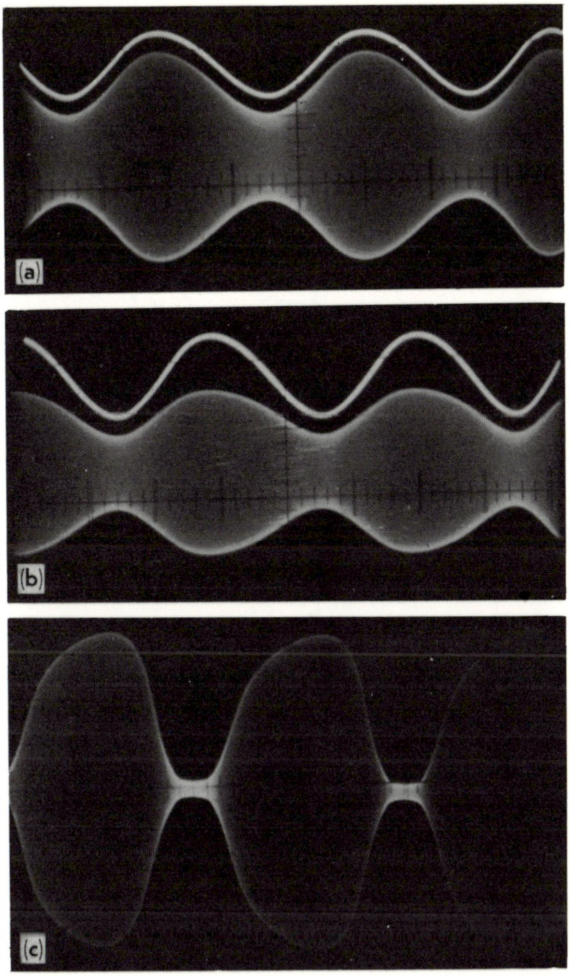

FIG. 7.8. *Modulation waveforms. At* (a) *is shown the modulation envelope on one trace and the modulation signal on the other trace; these reveal very little envelope distortion. At* (b) *severe envelope distortion is present since the carrier does not expand fully to follow the positive peaks of the modulation signal. At* (c) *is shown modulation signal distortion at high modulation feed.*

found by adding the maximum and minimum amplitudes. The modulation fraction is found by dividing the difference by the sum, and the percentage of modulation (or depth) is found by multiplying the modulation fraction by 100. The modulation percentage of the modulated waveform of Fig. 7.8*a* is approximately 45 per cent, and that of Fig. 7.8*b* approximately 33 per cent.

With 100 per cent modulation the carrier amplitude is reduced to zero during the negative half-cycles of modulation, and in Fig. 7.8*c* a distorted modulation envelope (due to a.f. distortion) is shown.

The application of modulation should cause the amplitude of the carrier to increase and decrease alternatively, but it sometimes happens that a fault in the modulation system causes only a *decrease* in carrier amplitude, as in Fig. 7.8*b*.

Under this condition the radiated power is considerably impaired since the carrier cannot expand as it should do. If the carrier amplitude expands to twice its normal value on positive peaks (as it should do on 100 per cent modulation), the peak radiated power is increased four times.

Another modulation pattern is shown in Fig. 7.9. This is obtained by applying the modulation signal to the X plates (with the timebase switched off) and the modulated r.f. to the Y plates or Y amplifier. With partial modulation the end

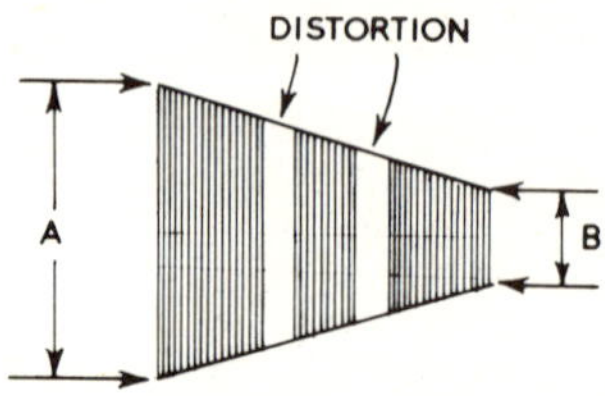

FIG. 7.9. *An alternative form of modulation pattern, from which the modulation percentage may also be calculated. Note the effect of distortion.*

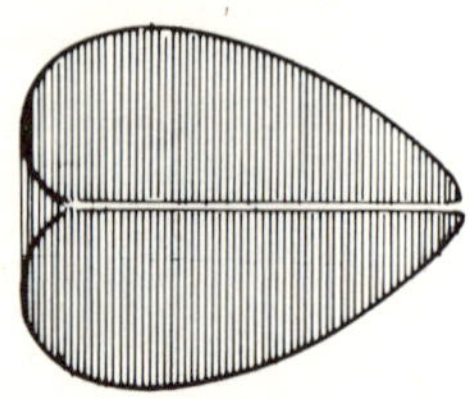

FIG. 7.10. *When the modulation pattern appears to be wrapped round a cylinder, serious phase distortion is present.*

of the trapezium pattern never reaches the zero line, whereas with over-modulation it falls to zero too soon and finishes in a straight line. The presence of distortion is revealed by vertical bands across the display, as shown in the diagram. A serious form of phase distortion is revealed on the display if the modulation trapezium appears as though wrapped round a cylinder. This is shown in Fig. 7.10. It should be noted that such phase distortion may sometimes be introduced in the Y amplifier (or X amplifier) of the oscilloscope.

Modulation depth may be calculated from the trapezium in a manner similar to that already described for the modulation envelope. In relation to Fig. 7.9, the percentage modulation is:

$$100(A-B)/(A+B)$$

If a modulated r.f. signal is applied between the aerial and earth terminals of a radio receiver or to the input of the i.f. amplifier distortion can be checked by observing the modulation envelope appearing at the detector input. The Y amplifier terminal of the oscilloscope should be connected to the detector diode through a 100pF capacitor, and the earth terminal should be connected to receiver chassis—taking any required precautions for isolating the chassis.

If there is no distortion a modulation envelope display identical to that at the output of the signal generator will be obtained. The amplitude will be greater, of course, but to avoid overloading the receiver stages and the Y amplifier only a small signal should be applied from the generator. If a double-beam oscilloscope is available, the generator signal can be applied to one beam and the signal at the detector input can be applied to the other beam. In this way the two modulation envelopes may be compared by adjusting the Y gain controls to provide displays of equal amplitude. By judicious adjustments to the gain and shift controls it is possible to superimpose one trace upon the other, and any slight deviation of the amplified waveform from the applied waveform becomes apparent.

On some oscilloscopes the Y amplifier response may not extend to the r.f. or i.f. range, in which case the Y plate may be connected direct to the detector input, via a 100pF capacitor. If there appears to be some form of distortion, the modulation should be switched off and the unmodulated carrier wave observed on the screen. The sweep frequency of the oscilloscope will have been set to suit the modulation frequency, so the carrier will appear as a blur on the screen because the separate waves will be too close together to be distinguishable. The separation between top and bottom is proportional to the peak-to-peak amplitude of the carrier.

R.F. or i.f. distortion can be discovered by examining the band of light, comprising the many r.f. or i.f. wave cycles. Fig. 7.11*a* shows a normal band of cycles, and Fig. 7.11*b* several cycles of a distorted r.f. waveform. When such distorted cycles are very close together in the form of a band of light (e.g., when the time-base frequency is low compared with the signal frequency), the distorted parts show up as light streaks horizontally across the display, as shown in Fig. 7.11*c*.

The reason for this is that the individual waves remain for a longer period at the distorted points (centre and bottom), with the result that bands of light

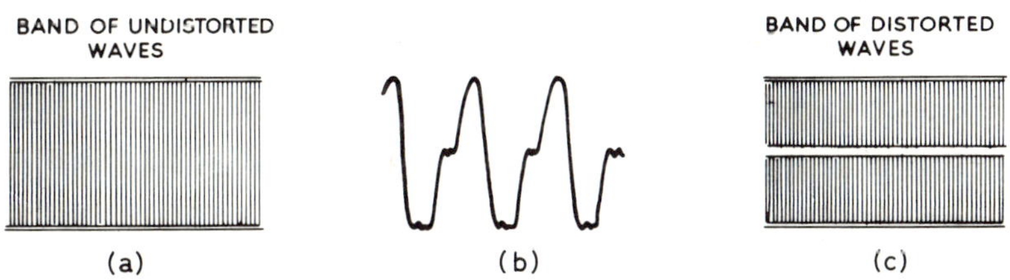

FIG. 7.11. *A band of waves as at* (a) *appears with horizontal white streaks as at* (c) *when each cycle is distorted as at* (b).

appear slightly brighter in the centre and at the bottom. The same applies to some extent with any sine wave; the top and bottom of a band of such cycles must be slightly brighter than the middle section because the spot of light is moving more slowly at these points. Where there is distortion, however, the white bands or streaks are more prominent than at the top and bottom of a band of sine waves.

If such distortion or any distortion is indicated on the r.f. or i.f. waveform band, the signal should be applied to the i.f. input. If distortion is still present, then it is undoubtedly in the i.f. amplifier; if it disappears, however, the operating conditions of the r.f. amplifier or frequency-changer should be investigated.

If the Y lead is now transferred to the live side of the detector load resistor and the generator modulation restored, the demodulated a.f. waveform will be displayed. This should exactly follow the contours of one side of the modulation envelope, indicating zero distortion. If a double-beam oscilloscope is available, the modulation envelope at the anode of the detector can be observed on one trace and the demodulated a.f. on the other trace giving a display similar to that in Fig. 7.8*a*. By adjusting the Y gain controls and the shift controls, the demodulated wave can be superimposed on the appropriate envelope of the modulated wave for direct comparison. Should there be any marked difference between the waveform contours, attention should be directed towards the operating conditions of the detector state; the diode, load resistor and associated capacitors should be checked for value.

It will be apparent that the oscilloscope can be used for numerous tests in radio, television and audio equipment. Faults which result in no output (e.g., when the oscilloscope is used to trace signals) and cause incorrect patterns and distortion, must always lie somewhere between the applied signal source and the point of connexion of the Y amplifier, and the object should always be that of reducing the amount of circuitry between the applied signal and the oscilloscope, until eventually only the faulty item remains.

DECOUPLING AND HUM

The oscilloscope can also be usefully employed for checking decoupling networks. For example, the screen grid of an i.f. amplifier valve is always decoupled to the i.f. signal by connecting the screen to chassis through a suitable value capacitor. The capacitor isolates the d.c. component, but acts as a short circuit to the i.f. signal. Thus, if the decoupling is working properly, there should be zero signal displayed when an oscilloscope is connected between the screen grid and chassis. The efficiency of the decoupling may be gleaned by trying this test both with the capacitor connected and disconnected.

The same procedure may be used to check the decoupling in transistor equipment. The oscilloscope must, of course, be able to respond to the frequency of the signals being decoupled, and should have a good Y amplifier to secure full benefit from such tests.

Hum voltages can be checked in a like manner by connecting the Y amplifier, via an 0·25-microfarad non-electrolytic capacitor to various points in the h.t. filter and feed circuits. The 'earth' terminal should be connected to chassis and the timebase should be set to give several cycles at the power frequency. First the Y lead should be connected to the cathode or d.c. output of the rectifier, where the hum is normally fairly high, and the Y gain should be adjusted to give about two-thirds vertical deflection. The Y lead should then be moved towards the h.t. line and to the various smoothed circuits. The hum voltage should become progressively less peaky and of smaller amplitude, possibly requiring greater Y gain in order to observe it at all.

The improvement in hum achieved by connecting a replacement smoothing or electrolytic capacitor can easily be observed. This will also indicate whether or not the existing component is in order. It should be noted that with half-wave rectification the hum voltage will be at the power frequency, and twice the power frequency with full-wave rectification.

The oscilloscope has many other applications that just cannot be dealt with here. However, quite a bit of information is given in Chapter 11 on the application of the instrument for testing and servicing in audio equipment, while reference is again made to the instrument in Chapter 10 for testing in colour receivers. Readers keen to obtain more information on the oscilloscope and its applications might care to refer to my companion book by the same publisher entitled *Servicing with the Oscilloscope.*

CHAPTER 8

Valve, Picture Tube and Transistor Testing

FROM the practical point of view, one of the most conclusive ways to test a valve, picture tube or transistor is to substitute the suspect item with a replacement known to be in good order. This method of checking is practised extensively by service technicians all over the world and has much to commend it, especially if the fault happens to be of the kind that may not show up on a tester and yet disturbs the normal operation of a critical circuit.

There are, however, several disadvantages to this method, apart from the obvious one of having to maintain a large stock of replacements. A direct replacement may not be available; a technician may be presented with several valves to test, without the equipment to which they belong, or definite measurements may be required by an experimenter or set owner. In all of these cases some kind of tester or test set-up is the only answer.

VALVE TESTING

One of the first things that can easily be found out about a valve is whether or not its filament or heater is intact. This can be done simply by connecting an ohmmeter across the filament or heater pins on the base. (Valve base connexions can be checked from valve manuals and similar publications.) Apart from the ohmmeter, however, there are several instruments available commercially which are designed with the chief application of making such continuity tests. Indication of continuity may be given on a meter or a neon bulb. When making this kind of test on the filament of a low-consumption battery valve, care should be taken to ensure that the current required to operate the meter or continuity tester does not exceed the current rating of the filament. If it does, then the test will either fuse the filament or possibly ruin the emission. This can happen with certain inexpensive types of low-sensitivity multirange meters having an 'ohms' range which reads only to a maximum of a thousand or so ohms.

If the filament or heater is found to be open-circuit, then, of course, there is little point in continuing with the testing.

A test of emission is usually the next requirement, and this requires the measurement of mutual conductance or 'slope' (called g_m and given in milliamperes per volt). That is the working principle of most valve testers: the valve under test is supplied with voltages approximating those experienced under normal operating conditions, and pre-determined variations of grid bias enables the g_m of the valve to be read from a calibrated milliammeter inserted to indicate anode current.

The slope of a valve is a change in anode current divided by the change in control grid voltage necessary to produce it. For example:

$$g_m = \frac{\text{Small change in anode current}}{\text{Small change in grid voltage}}$$

Thus, if 1 volt change in grid bias causes 6 mA change in anode current, the g_m is 6 mA/V, since it is the anode current change relative to a 1-volt change in grid voltage. This characteristic, like other valve characteristics, is based on the fact that all other voltages remain unchanged. Fig. 8.1 is a simple circuit which may be used to check mutual conductance. The switch simply changes the grid bias by 1 volt and the resulting difference in anode current is indicated by

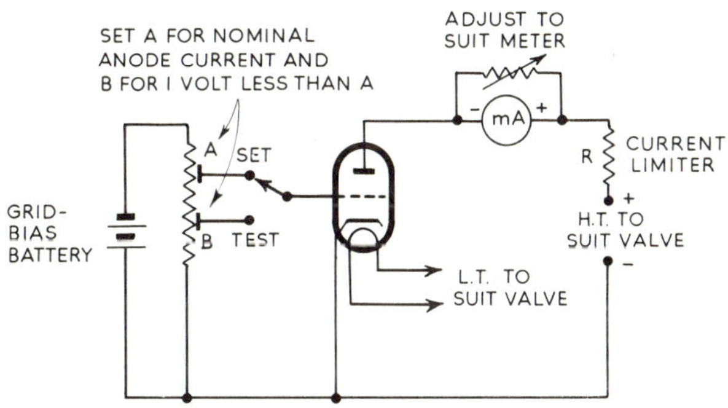

FIG. 8.1. *Simple set-up for checking mutual conductance. Multi-electrode valves may be checked in a like manner by connecting the appropriate electrodes to points of h.t. voltage.*

the milliammeter in the anode circuit. In valve testers a similar basic arrangement is sometimes adopted, but a bridge circuit and set-zero control enables the milliammeter to be set to zero on a calibrated mutual-conductance scale. When the test switch is operated, the grid bias is reduced by 1 volt and the milliammeter deflects to read the g_m direct in mA/V.

Mutual conductance is a factor which combines in one term the amplification factor (μ) and the anode resistance or impedance (r_a). Thus:

$$g_m = \mu / r_a$$
$$r_a = \mu / g_m$$
$$\mu = g_m \times r_a$$

where r_a is measured in thousands of ohms.

As a valve ages, its anode impedance increases and in consequence its mutual conductance decreases. It may be seen, therefore, that a measure of either g_m or r_a indicates the quality of a valve. In addition to being calibrated in g_m, some valve testers feature colour-graded scales, on which the valve gives a reading of 'good' or 'bad', as the case may be.

Mutual-conductance values are to be found in valve data books issued by most manufacturers, but it should be noted that the g_m quoted is that obtained under certain operating conditions which may not be similar to the test conditions. It is for this reason that a comprehensive test book giving the readings to be expected with the majority of valves, is supplied with valve testers.

Moreover, when a valve is set-up for a g_m test in accordance with the manufacturer's specifications, the exact g_m value or 100 per cent may not be given by the meter. New valves will sometimes show indications greater than 100 per cent. This is because valves are manufactured within a given tolerance, and certain valves are obviously going to be at the high end of the tolerance (sometimes called design centre), while others are going to be at the low end of the tolerance, thereby giving readings less than 100 per cent.

Other Tests

Older valve testers and certain modern inexpensive models simply give an indication of cathode emission. Their general arrangement is shown in Fig. 8.2. All the electrodes except cathode and heater are connected together and a

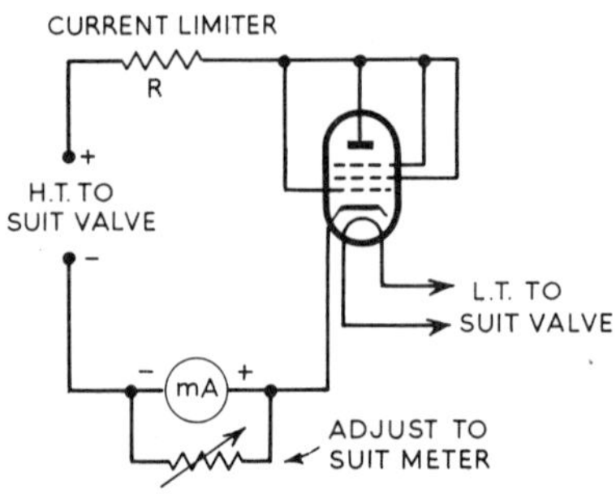

FIG. 8.2. *Circuit arrangement for checking cathode emission. This method is also used for checking diode valves.*

milliammeter is connected in the cathode circuit to measure the cathode current when the heater or filament is energised and the valve is connected across a high-tension source of relatively low voltage.

The milliammeter may feature a coloured scale giving such indications as 'bad', 'poor', 'replace', etc. and 'good' or 'normal' at the high-current end of the scale. There are slight variations of this arrangement, one of which is that the control grid, instead of being connected to anode, is connected to a small negative bias or even to a point of zero bias. In this way the potential on the control grid will influence the emission indication to some extent.

Apart from g_m and cathode emission, there are other tests required on valves. These include a short circuit or leakage between electrodes, grid current and microphony or noise. Short circuits between the electrodes can be easily checked by connecting a continuity meter or ohmmeter to the various pins. Poor insulation between the heater and cathode may, however, be impossible to assess by this means as the effect may occur only when the cathode is hot or after the valve has been operating under normal conditions for some length of time. The same may, in fact, apply to defective insulation between other electrodes. Microphony and noise, on the other hand, can usually be detected with the valve in the receiver by subjecting it to excessive vibration, by gently tapping it with the eraser end of a pencil, for example. Grid current, providing an indication of vacuum, can also only be checked successfully by some test set-up, or by measuring the grid current on a very sensitive meter when the valve is operating normally.

VALVE TESTERS

Although valve testers are in essence simple instruments, they are somewhat complicated by the required assembly of various types of valveholders (internally paralleled) and by the switching arrangements provided internally to apply the correct operating voltages to certain valveholders for many hundreds of different types of valves.

Modern valve testers provide such tests as continuity of the filament or heater; insulation between electrodes to manufacturers' specifications (when the valve is cold and also when it is in receipt of its various operating voltages); insulation between heater and cathode (hot and cold); that the vacuum of the valve remains at a suitable figure and that the reverse grid current, due to unwanted leakages or other conditions, is not excessive. (In this last test the total grid current, due to all causes, is usually measured and checked to discover whether it exceeds the maximum limit prescribed by the valve manufacturer.) A further test is the measurement of anode current under operating conditions; that is, with the normal operating voltages applied to each electrode, the anode current is measured to ascertain whether or not it is above the 'low' limit given by the manufacturer.

Other tests sometimes include a check to ensure that each electrode in the valve is properly connected to its appropriate valve pin, and a check for microphony and noise, which is carried out by the insertion of a pair of headphones, using a jack plug fitted to the instrument.

Most valve testers also check diodes and power rectifiers. In the Avo Valve Characteristic Meter, for example, power rectifiers may be checked under full-load conditions with a reservoir capacitor of 8 microfarads and with load currents of 5, 15, 30, 60 or 120 mA to each anode. The condition of the valve is indicated on a 'good-bad' scale. Ordinary diodes may be checked under reservoir-load conditions at 1 mA. This tester also enables a large number of valve characteristics to be accurately determined.

Indication of the condition of the test valve is primarily given on a moving-coil meter. However, this practice was modified by the Mullard High Speed Valve Tester which used a cathode-ray tube. The instrument was designed so that a non-technical person could learn to operate it in a very short time. It works on a 'test card' system, in which the appropriate card for the valve under test is inserted into a slot at the rear of the case. When a side lever is pulled down, a group of spring-loaded contacts press against the card and some make connexion with a further group of contacts through perforations in the card. Each card is coded and selects automatically the required 'test programme' for the type of valve to which it relates. In other words, 'computerised' valve testing.

POINTS ABOUT TESTING VALVES

A good valve will operate satisfactorily for a while on a reduced heater rating, but the emission of an old valve will quickly fade away. This fact provides the basis of a reliable guide to the condition of a valve which has already given service, and will enable the service technician or experimenter to decide which valves should be replaced during the general overhaul of a receiver. The valve under test should be allowed to warm up in the valve tester until its emission becomes stable at a figure within its normal rating. The heater voltage should then be reduced by 10 to 15 per cent and the emission figure noted when it reaches its new steady value.

If the valve has a good reserve of emission the change in anode current should not be more than about 15 to 25 per cent, depending on the type of valve. This test is not applicable to new valves because their emission is not always completely stabilised until they have seen some service.

When a valve operates for only a short time before giving distortion or low gain, the reason may be excessive grid current. If this current is caused by grid emission or grid-to-cathode leakage, the effect will not be evident until the valve has thoroughly warmed up, a condition which may not be produced in a valve tester, where there is plenty of ventilation around the glass envelope. In such cases the test may best be undertaken in a receiver by observing the change in cathode bias when the grid resistor is temporarily short-circuited.

If the vacuum has been destroyed by a cracked envelope or leaky seal, the heater may fail to light up because the air in the valve will prevent the heater from reaching its normal operating temperature.

Valves with partially shorting heaters, which are used in a constant-current heater-chain circuit of a.c./d.c. type receivers, will fail to light in the receiver but may show no fault in a tester owing to the constant-voltage heater supply used.

Battery valves may sometimes develop an intermittent, though possibly temporary, filament-to-grid short when operated in a valve tester which uses an a.c. supply for the filament. This is because the a.c. may cause the filament to vibrate at sufficient amplitude to make contact with the grid, especially if the valve is tapped or subjected to excessive vibration during the test.

PICTURE-TUBE TESTING

Of recent years picture-tube testers have been developed, and certain valve testers have been adapted to undertake one or two tests on picture tubes.

One of the main tests is for cathode emission. The heater is energised in the normal manner and the cathode emission is assessed by a milliammeter connected to the grid which is arranged to act as an anode by being connected to a source of voltage which is positive with respect to the cathode. The general idea is shown in Fig. 8.3.

Other tests include filament continuity, inter-electrode insulation and, in some cases, final-anode beam current. Some testers like this have been made

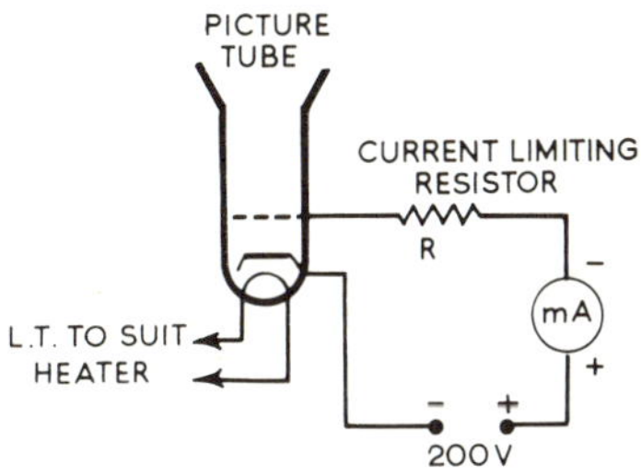

FIG. 8.3. *Circuit arrangement for checking the cathode emission of a picture tube. Comparative measurements are necessary on tubes of similar type to obtain reference data.*

also to include facilities for reactivating worn tubes and for clearing inter-electrode shorts.

In addition to checking heater continuity, some instruments measure the heater current on two ranges and thus speedily reveal whether or not the tube has a partially shorting heater. Reactivation by the instrument shown in Fig. 8.4 is controlled by a four-position switch which gives varying degrees of processing by a combination of pulsed cathode emission and, in the case of very weak tubes, heater boosting. Inter-electrode shorts may be cleared by a 'clear leak' device. With this sort of instrument there is no need to remove the tube from the receiver, as connexion from the instrument to the tube is via a flexible

FIG. 8.4. *Example of Tube Tester and Reactivator.*

lead and universal socket unit which will accommodate all types of tubes.

In addition to its function as a picture-tube tester and reactivator, the instrument shown contains facilities to check the potentials applied to the tube while operating in the receiver. Ranges measure the operating heater voltage (0–25 volts a.c.), grid, cathode and final-anode potentials (0–500 volts d.c. and 0–25 kV d.c.).

In Situ Tests

With the necessary instruments *in situ* tests can be made on a monochrome (not colour) picture tube with a high degree of success. However, before making any measurements, it is particularly important to ensure that the ion trap magnet is in its correct position on the neck of the tube, that the external conductive coating, if employed, is effectively connected to the receiver chassis, and that the heater is glowing at the correct brightness.

The heater current or voltage should first be measured on an a.c. meter close to the tube heater pins, so that any voltage drop in the heater wires is indicated. It should be noted that high-current (low-voltage) tubes are particularly vulnerable to voltage drop in the leads, and that this can impair the life of the tube.

Across the heater of a tube operating in a series heater chain, a low voltage yet correct current would indicate a partial short in the heater.

The final-anode e.h.t. voltage should be checked on a suitable meter.

The first-anode h.t. voltage should be checked on a very-high-resistance meter (preferably a valve voltmeter) owing to the high value of the feed resistor used in most sets. The measurement should be taken between the first anode and the *cathode*—not the chassis.

The grid voltage may be checked on a high-resistance meter connected between the control grid and cathode pins of the socket. Variation of the brightness control should vary the meter reading. If no raster is obtained when the grid voltage is reduced to approximately −5 volts, a measurement of cathode current should be made. This is accomplished by inserting a microammeter in series with the lead to the cathode (with the set switched off). After switching on, the grid voltage should be adjusted according to the figures given in the makers' data sheet to give the required beam current (20 to 30 microamperes). If current of this order is present but no raster appears on the screen, the ion trap magnet should be very carefully adjusted and if necessary changed for a new one. It should be noted that the actual value of grid bias required for a certain beam current varies widely from tube to tube.

Uncontrollable brightness or poor picture definition (of which the latter may be intermittently corrected by tapping the neck of the tube) may be caused by poor insulation between heater and cathode. This can usually be detected by removing the tube base and connecting an ohmmeter between the heater and cathode pins. As with valves, this defect may only occur when the cathode is hot.

It is generally undesirable to attempt such tests on colour tubes.

TESTING TRANSISTORS

A transistor may be likened to a pair of semiconductor diodes integrated in a single unit. In operation, one of these diodes is given a 'reverse' bias and the other a 'forward' bias. The diode in 'forward' bias is that represented between the emitter and base terminals or wires of the transistor, while the diode in 'reverse' bias is that represented between the collector and base terminals. The base element is thus common in each case.

With this in mind, some useful information can be obtained on the goodness or otherwise of a transistor by nothing more elaborate than an ohmmeter. If each diode section is in order, then different values of resistance will be registered depending on the polarity of the applied voltage. The voltage is derived from the battery contained inside the ohmmeter and this, of course, appears across the test leads.

The test is made first by applying the test leads between the base and emitter terminals and noting the resistance reading, and then reversing the test leads and comparing the resistance reading now obtained with the first reading. A low resistance reading should be obtained in one direction (when the polarity of the voltage applied to the elements is such that the diode is in forward bias) and a high resistance in the other direction (when the diode is in reverse bias).

The test should then be repeated by measuring the resistance in both directions across the base and collector terminals, again comparing the readings.

With low-power transistors the resistance in forward conduction may be less than 500 ohms, while the resistance in reverse conduction may be greater than 10,000 to 15,000 ohms. If the resistance in both forward and reverse conduction is low, with no difference between the two readings, the transistor is short-circuited or very leaky across that particular diode section. If there is no difference between the two readings, whether the actual values are high or low, the transistor is faulty, and a very high resistance—or no reading at all—would indicate an open-circuited diode junction.

The actual resistance values will depend on the type of transistor under test. Power transistors usually have a very low forward resistance, but in all cases the ratio between the two values should remain high—at least 25 to 1 and as high as 100 to 1 or more on high-quality components.

This kind of test can also be used on semiconductor diodes, but both with diodes and transistors care must be taken to avoid overrunning the device with excessive current. A series-type ohmmeter should always be employed and, if necessary, a limiting resistor should be connected in series to hold forward currents to safe values.

TEST CIRCUITS FOR TRANSISTORS

Fig. 8.5 shows some transistor test circuits. At (*a*) leakage current in the collector circuit ($I'co$) and collector current at cut-off with the base connected to emitter may be measured. Switch *S* provides the two tests and the readings may be taken on a microammeter over a range of collector voltages.

At (*b*) is given a circuit for measuring the collector turnover voltage (V_t). The collector will normally pass a small leakage current up to a certain collector voltage. At this point the reverse current in the collector-base junction rises sharply to a high value as the voltage is further increased. This effect is also exhibited in semiconductor diodes, and is known as the 'reverse turnover voltage'. Excessive reverse current can destroy the diode or transistor, and for this reason a current-limiting resistor should be included in the test circuit as shown. The turnover point will be indicated by the voltage across the transistor falling rapidly owing to the voltage drop in the limiting resistor.

At (*c*) is given a test set-up which in addition to measuring leakage current, will measure current gain. The circuit is powered from a 9-volt grid-bias battery, and a tapping for the collector is made at the 1·5-volt point and led via a milliammeter. The current gain (β) test measures a change of collector current caused by a change in base current, which corresponds to the change in anode current for a change in grid voltage in a thermionic valve. [The parameter h_{FE} is now preferred (in common-emitter mode) to β.]

This test is carried out first by adjusting the base current for a normal static bias condition by the variable resistor P_1. This should provide a centre reading on the milliammeter, whose full-scale value should be of the order of 2 mA.

Resistor *R*, in series with press-switch S_2, allows the base current to be increased by an amount governed by the value of *R*. This extra current, of course,

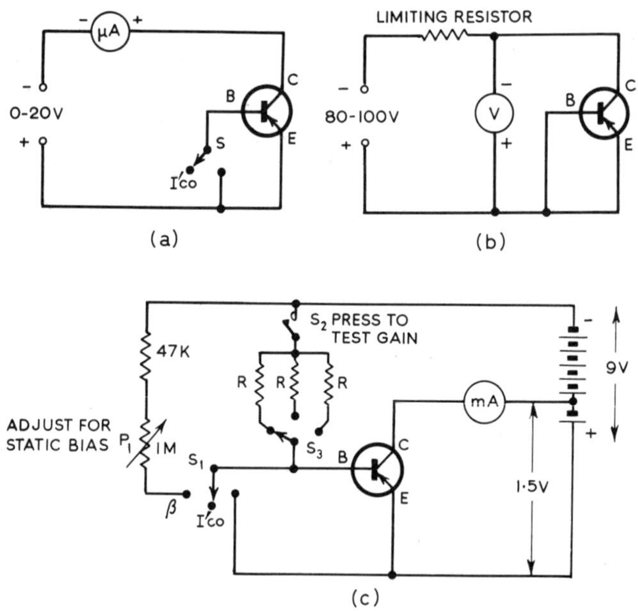

FIG. 8.5. (a) *Set-up for measuring leakage current with earthed emitter and with earthed base.* (b) *Set-up for measuring collector turnover voltage (V_t).* (c) *Test circuit which also measures current gain (β).*

FIG. 8.6. *Avo Transistor Tester Type TT166. This is intended for the rapid measurement of the d.c. parameters of bipolar transistors and is capable of measuring current gain at very low collector currents. Current gain can be measured at four values of collector current over the range 1 μA to 1 mA on both npn and pnp devices. There are also facilities for the measurement of emitter/base voltage and collector/base leakage current at any one of the four collector currents. Measurement can also be made of reverse leakage current and forward conduction voltage of semiconductor diodes. The instrument works from two internal 9 V batteries.*

is applied in parallel with the static bias current as set by P_1. The value for R is best calculated in relation to a transistor whose current gain is, say, 100, and to provide full-scale deflection on the meter under these conditions. If the meter has 2 mA full-scale deflection, and is set to 1 mA for the static condition, then the remaining 1-mA deflection should occur when the current gain equals 100. Thus, as the resulting change in collector current is related to the current gain factor:

$$\text{Current gain} = dI_c/dI_b$$

where the symbol d indicates change, I_c is collector current and I_b is base current, it follows that with a transistor with a gain factor of 100 a change of 1 mA will result from a base current increase in milliamperes of 1/100, which works out to 0·01 mA or 10 microamperes.

Therefore, in order to obtain full-scale deflection on the 2-mA meter, indicating a current gain of 100, the resistor R must feed to the base on operation of the press-switch an extra 0·01 mA. The value for R may thus be obtained from Ohm's law (e.g., $R = E/I = 9/0{\cdot}00001$), and in this case works out to 900,000 ohms. To avoid error when measuring relatively low gain transistors, the full-scale deflection, taking into consideration the static current, may be adjusted for gains of, say 25, 50, 75 and 100, by switching in suitable values for

R. Alternatively, the second half of the meter scale may be calibrated direct in terms of gain factors.

The switched *R* arrangement is shown in the diagram, operated by switch S_3. The specified gain of the transistor is set on S_3, and if the transistor is fully up to standard a full-scale reading on the meter should be obtained when S_2 is depressed. Plus and minus tolerance must, of course, be expected in practice, as with valves.

Most transistors now in service in Britain are npn types, and these differ from pnp types by requiring potentials of opposite polarity. It is important, therefore, to ascertain the type of transistor under test, for a reversed voltage connexion may impair the transistor.

Transistors are also affected by temperature. For example, the small current which flows in the collector with zero base current varies with temperature, and the greater this leakage current the greater is the change in current with change in temperature. In practice, various methods are adopted for stabilising this effect.

Transistor Parameters

There is still some lack of standardisation with regard to certain transistor parameters, though many have become established by common usage. For example, h_{FE} is now commonly accepted as the forward current-transfer ratio in the common-emitter mode, while h_{fe} is the small-signal equivalent I_{CBO} now denotes the collector cut-off current with the emitter open-circuit and I_{ceo} with the base open-circuit.

h_{FB} is the forward current-transfer ratio in the common-base mode, with h_{fb} denoting the small-signal equivalent. Other parameters include I_{ebo}, which is the emitter cut-off current with the collector open-circuit, and f_T, which is the gain-bandwidth product measured in the common-emitter mode.

CHAPTER 9

Miscellaneous Instruments

THERE ARE hosts of instruments available to the service technician, experimenter and electronics engineer which have somewhat specialised applications. There are others, however, which, although not in the category of basic instruments, are of a less specialised nature and have reasonable scope for application in modern servicing. Some of these are considered in this chapter.

SIGNAL STRENGTH METER

Before the days of television or v.h.f. radio a signal strength meter (sometimes called a 'field strength' meter) was a highly complex and expensive piece of equipment, for which there was very little call. Today the situation is entirely different. Now that untuned long-wire aerials have given place to tuned dipoles and highly directional arrays, a method is necessary to ascertain the approximate strength of the signal induced into the aerial and fed to the receiver through low-impedance coaxial cable. This is necessary firstly to ensure that the array is orientated for maximum signal pick-up, and secondly that an apparent fault in a receiver is not caused by insufficient signal.

With v.h.f. signals, as compared to medium-frequency signals, there are many more areas where the signal strength is below the required standard. This may be caused by distance from a transmitter, or by local screening effects; in both of these cases an aerial of special type, or extra gain, may be required in order to obtain the maximum pick-up of signal. Some means of measuring or comparing signals is thus desirable and almost essential in some instances. Hoping to get the display itself to detect changes in signal level does not work these days, since the automatic gain control system, on both sound and vision, is extremely efficient in making signal variation, and it is very difficult in practice to adjust the aerial system for minimum noise.

Considered in this light, a signal strength meter is as important to the modern service technician as, for example, a multirange testmeter. It is also invaluable to the experimenter interested in constructing and comparing the performance of aerial systems.

A signal strength meter measures the signal voltage which actually exists across the end of a downlead, and is usually calibrated directly in millivolts or microvolts. It is designed so that its input represents a reasonable match to the characteristic impedance of the coaxial downlead. The instrument must, of course, be tunable over the bands in which it will be required to measure signal strengths, since it must respond to only one signal at a time. There are several instruments of this nature available commercially, and most of them feature the same basic principles of operation.

The front end looks very much like an ordinary radio receiver. There is usually a frequency-changer stage preceded by an r.f. amplifier, and these stages are tuned either by a ganged capacitor, in conjunction with a calibrated tuning dial, or by a conventional turret or switched tuner of the type used in television receivers.

The tuner section then feeds into an i.f. amplifier stage, and this feeds a detector or rectifier. Instead of the rectified signal being used for driving a video or a.f. amplifier, it is used as a control bias for the i.f. amplifier valve. In this valve's anode circuit (or cathode circuit) is a milliammeter, and as the control bias increases, as would result from an increase in signal strength, so the current in the i.f. valve decreases and thus causes a deflection on the milliammeter.

Under no-signal-input conditions, the standing bias on the i.f. valve is adjusted to give full-scale deflection on the meter (a 'set zero' control being available for this purpose) which corresponds to the zero-signal calibration mark on the scale.

When a signal is applied, the deflection on the meter decreases, and as the extent of the deflection—from maximum current to minimum current—is related to the strength of the applied signal, the scale can be calibrated direct in terms of signal voltage. A progressive cramping of the calibration occurs towards the high-signal (low-current) end of the scale owing to the non-linear

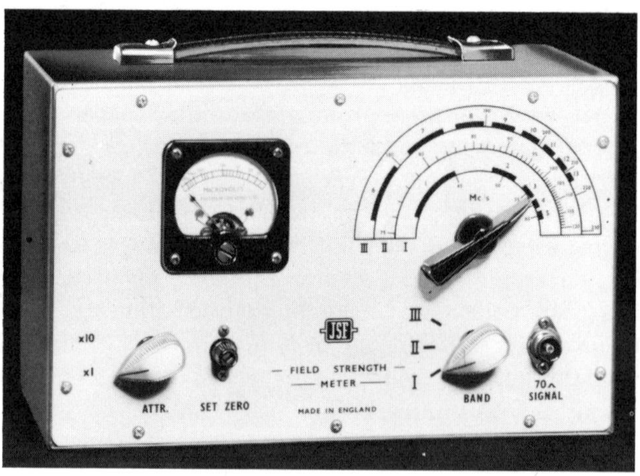

FIG. 9.1. *Example Field Strength Meter (J. S. Fielden)*

characteristics of the valve towards anode current cut-off. This is of little consequence, however, and may in some cases be desirable, since it ensures that signals in the microvolt region are calibrated over a larger section of the dial (where there is relative expansion of the calibration) and thus makes small signal strengths easier to measure.

It is also interesting that as increasing signal levels produce a decrease in the current through the meter, automatic protection to the meter against accidental overload is afforded. Battery or mains operated instruments are available, some suitable for both.

Among the applications of the signal strength meter are the following: installing and checking aerials on sites where a mains supply is not available; locating the most suitable spot for aerial installations; investigating type of aerial or receiver required for new installations; ensuring correct orientation of directional aerials for maximum signal pick-up; measuring signal strength; checking signal levels in coaxial relay and extended aerial systems; balancing Band I/III combined inputs; checks on aerials when investigating 'poor picture' service jobs; interference tracing and measurements; checking attenuators; and monitoring coaxial relay systems and transmitters.

Recent models also measure signals in the u.h.f. channels of Bands IV and V.

GRID-DIP OSCILLATOR

The grid-dip oscillator or wavemeter is simply a calibrated oscillator with a meter which reads grid current. Apart from its normal use as an oscillator, such an instrument can be used to check the resonant frequency of passive networks by observing the effect on the grid current when the circuit under test is coupled to the oscillator circuit.

In a normal oscillator circuit the amplitude of the oscillatory signal can be judged either from the grid current value or from the value of the current in the anode circuit by comparing this with the current when the valve is not oscillating.

The grid current indication is normally used, however, and this will drop almost to zero when the oscillation ceases. If the amplitude of oscillation decreases, then the grid current reading will decrease. If some of the energy of oscillation is coupled from the oscillator to a passive resonant circuit which is tuned to the same frequency as the oscillation, the amplitude of the oscillation will decrease and the grid reading will dip—hence the term 'grid-dip oscillator'.

The instrument proper contains the valve and components, including a calibrated tuning dial, that are required to produce the oscillation, while the coil is usually a plug-in component mounted at the end of the instrument in such a way that coupling between this coil and the tuned circuit under test can easily be accomplished. There is generally a set of coils which covers a wide frequency range.

Let us suppose that information on the resonant frequency of a parallel-tuned *LC* circuit is required (such as in the anode circuit of an r.f. amplifier). The coil of the grid-dip oscillator is held fairly close to the inductor of the external circuit, and the tuning of the instrument is carefully adjusted until a definite dip occurs in the meter reading. The frequency is then read straight from the dial or a graph.

If the coupling is too tight, the loading of the oscillator may be excessive and oscillation may cease. Too little coupling, on the other hand, will cause only a small dip on the meter which may be difficult to perceive and will possibly occur very quickly as the tuning is adjusted, thereby making the adjustment extremely sharp. Nevertheless, loose coupling gives the most accurate frequency indication. It is best to obtain first a rough idea of the frequency by using tight coupling, and then reduce the coupling to obtain a more accurate measurement. As the coupling is reduced the frequency will be found to alter slightly.

The frequency at resonance of any network may thus be measured, and circuits can be checked or pre-tuned before being incorporated into a unit or before power is applied to the circuits.

Such an instrument is extremely useful to the experimenter, for it enables him to wind coils and set up tuned circuits with a high degree of accuracy. If the equipment does not work on test, he can at least be certain that the tuned circuits are in adjustment.

By coupling the instrument's pick-up coil via a small loop to the downlead of a v.h.f. or television aerial, the resonant frequency of the aerial array can be measured. For experimenters who are keen on making and testing aerials, the grip-dip oscillator is invaluable. It will be appreciated, of course, that there are many more applications to which this versatile instrument may be put.

Recent versions employ solid-state devices, though the original name might still be used.

Measuring Apparent Inductance

Since the product of inductance and capacitance remains unaltered for any given frequency (e.g., if the inductance is *increased* then the capacitance must be *decreased* so that the *product* remains the same in order to hold the original resonant frequency, and vice versa), the grid-dip oscillator may be utilised to determine an unknown inductance, or apparent inductance. 'Apparent' because distributed capacitance exists across all coils due to the windings, etc.

A coil of known inductance (L_k) should be set up in parallel with a capacitor (C_k) of known value, and the grid-dip oscillator should be carefully tuned to the resonant frequency of the circuit. The unknown inductance (L_u) should then be connected in place of the known one and, without touching the tuning of the grid-dip oscillator, the capacitor should be altered in value until the circuit again comes into resonance at the original frequency. The new capacitance (C_n) should be noted, and the unknown inductance may be found from:

$$L_u = L_k \times C_k/C_n$$

The unknown inductance will be in the same units as the known inductance.

Measuring Distributed Capacitance

If, after making a test of inductance, as above, at another frequency, L_u is again computed it will be found to differ from the original. This is caused by distributed capacitance as already mentioned. This difference, however, may be used to measure the distributed capacitance. If the apparent inductance (L_u) is taken first at one frequency to give L_{u1} with C_{k1} and then at a lower frequency to give L_{u2} with C_{k2}, then the distributed capacitance (C_d) may be found as follows:

$$C_d = (L_{u1}/L_{u2} - 1) \times C_{k2} \times C_{k1}/C_{k2} - C_{k1}$$

where C_d is in the same units as C_{k1} and C_{k2}.

Grid-Dip Oscillator As Wavemeter

Most commercial grip-dip oscillators also act as wavemeters. As a wavemeter, the oscillator is switched off and the pick-up-coil is used to absorb energy from some other circuit in which there is a signal. For example, the pick-up coil may be held in proximity to the local-oscillator coil of a radio or television receiver. When the pick-up coil is tuned to the same frequency as the signal in the external circuit signal energy will pass from one to the other, and resonance will be indicated again on the meter, but this time it is, in effect, connected across the pick-up coil.

In this way the frequency of any fairly large amplitude signal may be determined. Some instruments have the oscillator valve arranged as a leaky-grid detector and incorporate a socket for headphones. Thus, when the instrument is tuned to the same frequency as the signal in an external circuit a heterodyne whistle will be heard in the headphones, and the frequency of the external signal may be accurately ascertained.

OSCILLOSCOPE TRACE DOUBLER

Double-beam oscilloscopes have the advantage that two waveforms may easily be compared. For example, when checking for distortion in an a.f. amplifier the actual waveform of the signal used to feed into the amplifier may be displayed on one beam and the waveform of the signal as it comes out of the amplifier may be displayed on the other beam. The two traces may then be superimposed and it is a simple matter to observe where one differs from the other.

It is not generally known that this facility can be provided on a single-beam oscilloscope by means of an electronic switch and amplifiers. Two amplifiers are employed to accept the two Y signals, and the outputs of the amplifiers are combined and fed to the Y plate of the oscilloscope. However, both amplifiers are not working simultaneously, since a square-wave gating or switching arrangement is also employed so that the amplifiers are switched on and off alternately. Each signal is, therefore, sampled at regular intervals governed by the switching rate. For example, the Y1 signal would go through the Y1 amplifier while the Y2 amplifier is switched off and give a small section of trace on the screen. The Y1 amplifier would then switch off and the Y2 signal would go through the Y2 amplifier which at that time would switch on, and a small section of the Y2 signal would be displayed on the screen.

The switch also allows for one trace section to be lifted above the other trace section, thereby permitting two separate traces to be displayed. It will be

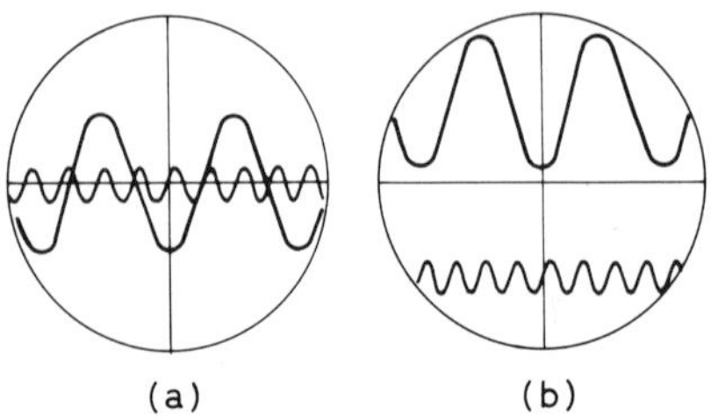

FIG. 9.2. *By the use of an electronic switch, a single-beam oscilloscope allows the display of twin traces to be made, which may be superimposed as at* (a) *or separated as at* (b).

appreciated that each trace is really made up of a series of dashes as the particular amplifier is switched on, but when the switching rate is fairly high each trace appears as a complete line, as shown in Fig. 9.2*a* and (*b*).

SIGNAL INJECTION PROBE

Details of signal tracing have already been given in relation to various pieces of equipment. Another 'signal tracing' technique which has recently been developed is a small instrument in the form of a probe which contains a small generator or oscillator, sometimes transistorised, producing square-wave signals rich in harmonics. By applying such a probe to the various stages of a receiver, from the output stage and working back towards the aerial, the square-wave signal will be heard from the loudspeaker up to the point where the circuit is defective. Normal servicing techniques can then be adopted to clear the trouble.

It will be understood that the square-wave signal provides a fundamental frequency in the audio spectrum, and this, in effect, is modulated on the higher-order harmonics, allowing the probe to be used in r.f. as well as a.f. applications.

MEASURING BRIDGES

The basic bridge circuit is shown in Fig. 9.4, and the bridge is balanced when

$$R_1/R_2 = R_3/R_4$$

or when

$$R_1 \times R_4 = R_2 \times R_3$$

Balance is indicated by the voltmeter indicating zero volts. If any arm is altered in value then the balance will be upset and the voltmeter will give a reading. Balance on a bridge is accomplished by adjusting the values of the arms for zero or minimum reading on the indicator, such as the voltmeter in Fig. 9.4.

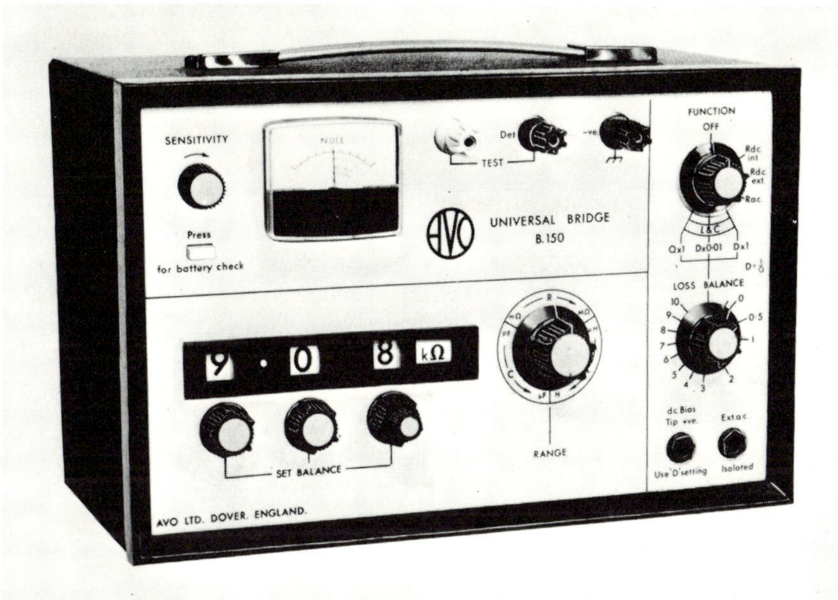

FIG. 9.3. *Avo Universal Measuring Bridge Type B150. This is transistorised and battery operated and has a modern digital in-line display. When the bridge is balanced, the digits and range symbols, indicating the component value, are displayed automatically. An inbuilt 1 kHz oscillator provides a wide range of L, C and R measurements; but measurements can be made at any other audio frequency by the use of an external oscillator.*

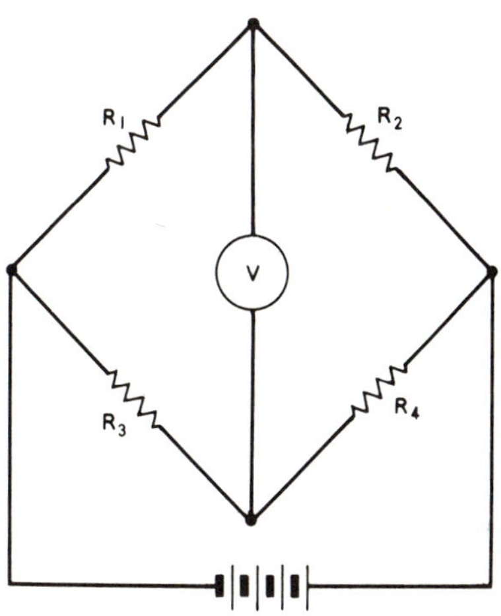

FIG. 9.4. *A basic Wheatstone bridge circuit. The voltmeter indicates zero when the bridge is balanced.*

In order to measure, say, resistance on a bridge, one arm of the bridge is represented by the unknown resistance (say R_4) and another arm is represented by a standard resistance (say R_3). R_1 and R_2 then serve as 'ratio arms' and the unknown resistance can be found as follows:

$$R_4 = R_3 \times R_2/R_1$$

Any arm may be taken as the standard but an adjacent arm must then be taken as the unknown. This leaves two adjacent resistances, which are the ratio arms. When built into a measuring bridge, the ratio arms are usually switched to give a range of values, and the standard is often in the form of a calibrated variable resistor. The indicating meter usually has a centre zero so that an accurate balance can be achieved, for if the balance is out one way a positive indication will be given and if out in the opposite way a negative indication will be given.

If R_1 (Fig. 9.4) is switched, say, to 100 ohms and R_2 to 300 ohms, then the ratio of the ratio arms is 1:3. If an unknown resistance is put in at R_4 and balance occurs when the standard R_3 is adjusted, say, to 600 ohms, then the unknown resistance is equal to one-third of 600 ohms, which is 200 ohms.

The same basic principle is adopted for the measurement of capacitance and inductance, but instead of d.c. being used a.c. is applied to the bridge, either at fairly low frequency for medium value capacitance and inductance, or at r.f. for very low value capacitance and inductance.

Where the source frequency is within the audio range, the indicating device may be a pair of headphones and the balancing element is adjusted for minimum sound from the headphones. An amplifier may be used between the bridge output and the headphones to facilitate adjusting to the null point. Alternatively, an audio millivoltmeter may be connected to the output of the bridge, the bridge then being adjusted in the same manner as when a voltmeter is employed in a d.c. resistive bridge. The millivoltmeter must, of course, respond to the frequency at which the bridge measurements are undertaken.

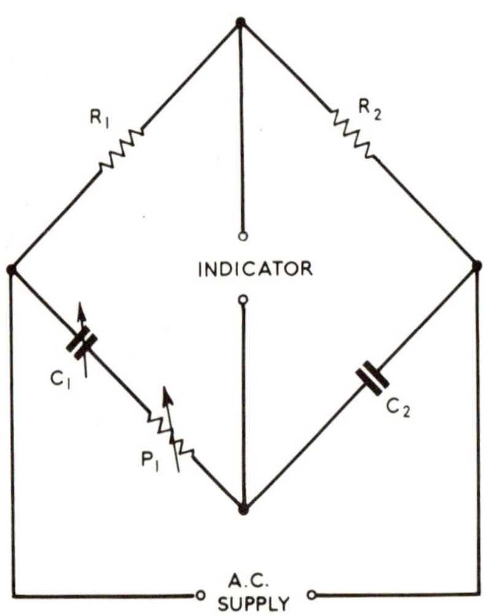

FIG. 9.5. *Basic bridge circuit for measuring capacitance. In an actual instrument the values of arms* R_1 *and* R_2 *are usually adjustable, by switching or other means, to adapt the instrument to a wide range of measurement.*

In Fig. 9.5 is shown the basic circuit usually employed for the measurement of capacitance. The bridge operates in terms of capacitive reactance or impedance (the latter is more accurate owing to the resistive component present in all capacitors). The capacitive reactance or impedance, which is measured in the same units as resistance, is then translated back to capacitance.

The ratio arms comprise resistances R_1 and R_2 and, in line with the resistance bridge, may be switched or adjustable to provide a wide range of measurements. C_1 is a standard variable capacitor, the dial of which is calibrated, while C_2 is the unknown capacitance. The variable resistor P_1 serves to balance out the resistive component of the unknown capacitance. This control is sometimes calibrated, particularly when the bridge is used to measure electrolytic capacitors, in terms of power factor, and is adjusted until the residual reading on the balance indicator reduces to absolute zero. The value of the unknown capacitance, in relation to Fig. 9.5, can be found from

$$C_2 = C_1 \times R_1/R_2$$

where C_2 is in the same units as C_1.

The basic bridge circuit for measuring inductance is shown in Fig. 9.6. The principle of operation is identical to that of the capacitance bridge and the

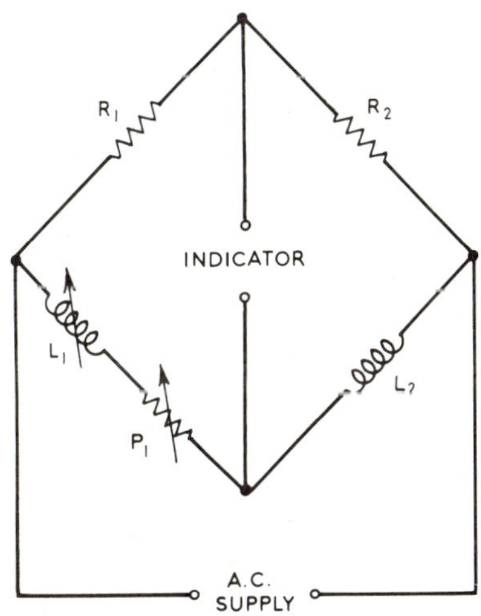

FIG. 9.6. *Basic bridge circuit for measuring inductance.*

bridge operates on the basis of inductive reactance or impedance. The value of the unknown inductance can be found from

$$L_2 = L_1 \times R_1/R_2$$

where L_2 is in the same units as L_1.

There are several variations of the bridges described, and some use fixed capacitors and inductors for standards instead of calibrated variable components. Bridge balance is then accomplished on a calibrated variable resistor. In most universal bridges, switching arrangements allow the measurement of resistance, inductance and capacitance over a very wide range of values.

Q METER

The Q meter (sometimes called 'circuit magnification meter') is a precision instrument rarely found in service departments or in experimenters' dens. It is, however, indispensable in development laboratories and in those places where it is necessary to obtain detailed information on coils and inductors. The Q meter allows the measurement of the Q values of coils over a variable frequency range; it indicates the resonant frequency of tuned circuits and also measures small inductance values; it measures the value of small capacitors and the self-capacitance of inductors.

'Q' is the factor symbolising 'quality' of an inductor, capacitor or tuned circuit. The greater the Q value, the greater the magnification or voltage step-up between that applied to the resonant circuit and that developed across either the inductor or the capacitor.

Inductors are endowed with a series resistance which, in addition to the d.c. resistance, takes into account the 'skin effect' of the wire at radio frequencies. Owing to the skin effect, the effective series resistance increases with increase in frequency. The expression for Q is

$$Q = 2\pi fL/R$$

where f is the frequency of the applied signal, L is the value of the inductance and R is the series resistance.

From this it will be seen that the Q value will be highest when L is large compared to R. High Q values are often desirable and, indeed, essential to provide high selectivity and sensitivity, and for this reason high-Q coils are often made with relatively thick wire (to reduce the resistance and skin effect) or with low-loss iron dust or ferrite cores to maintain the required inductance with less wire.

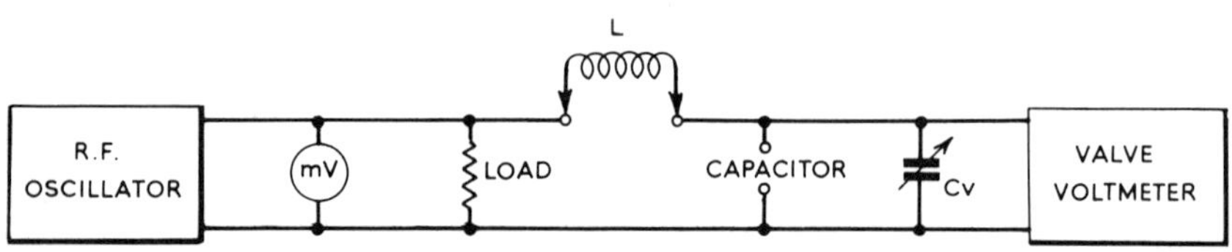

FIG. 9.7. *Essential requirements of a Q meter.*

When measuring Q, a signal of known level is applied to a series-tuned circuit comprising a variable capacitor and the inductor which is under test. A valve-voltmeter is then used to measure the voltage across the capacitor, and the step-up in voltage between that applied from the r.f. oscillator and that across the capacitor is equal to the Q value. It is, of course, necessary to bring the series-tuned circuit into resonance at the frequency at which it is required to measure the Q. The general idea is shown in Fig. 9.7, and the versatility of this set-up enables several secondary measurements to be effected.

For example, small values of capacitance may be measured. This is accomplished by adjusting the calibrated C_v to maximum value, inserting a suitable inductor and then adjusting the r.f. oscillator for resonance. The unknown capacitance is connected across the terminals marked 'capacitor' and C_v is again adjusted for resonance, at which point the capacitance value is noted. The value of the unknown capacitance is equal to the difference in capacitance as noted on V_c. Stray circuit capacitances may be measured in a like manner.

There are many more measurements for which the Q meter may be used, but these are beyond the scope of this book.

CHAPTER 10

Instruments for Colour Television

SINCE the advent of colour television it has been necessary for the service technician to acquire additional test equipment and possibly upgrade some of his existing equipment hitherto employed for the servicing of monochrome receivers and the various other items of domestic electronics which pass through his hands. It is the purpose of this chapter to survey those instruments designed specifically for colour television servicing, to discuss their uses, to discover the essential additions and upgradings and to provide some guidance on instrument choice.

At the outset, however, it must be emphasised that some items of colour receiver servicing equipment are suitable only for that purpose, while other items can be brought into operation for the servicing of equipment other than colour receivers. Moreover, colour receivers demand a more accurate home-setting-up procedure than their monochrome counterparts and the instruments required for this are sometimes less exacting than those used for more complicated workshop servicing. Nevertheless, colour receivers are generally heavier and bulkier than monochrome ones, and because the design trend is towards the use of plug-in printed circuit boards and 'modules' it is likely that a greater proportion of fault diagnosing and servicing work (since this may amount merely to locating the faulty printed circuit board or 'module' and fitting a replacement) will in future be done in the home.

The craft here, though, lies in diagnosing the trouble, so the skilled technician whose job this is is commonly equipped with test equipment of more exacting kind than the installation technician whose function is to deliver and satisfactorily establish the colour set in the viewer's home. Portable equipment of high standard, probably equal to that found in the servicing workshop, is fast becoming available to aid the diagnostic work of the service technician. The set makers are helping, too, by providing a complete range of replacement 'modules', etc. and well produced service manuals showing vulnerable test points, voltage values and waveforms.

All this boils down to the fact that colour test equipment can be considered in two parts. One, that required for installation and initial setting-up; and two, that for fault diagnosis both in the field and in the workshop. Let us look at the first part first.

One would normally assume that the colour set leaving the shop is in good working order which means that in theory it should work at the customer's home. In practice, however, some adjustments are required to provide optimum performance in the domestic environment not only due to the environmental change but also because vibrations in transit can impair some of the more 'delicate' adjustments performed at the shop. This happens to a smaller degree with monochrome sets, but the single-gun tube and the fewer presets generally allow the initial setting-up to be carried out on a test card without the help of instruments.

To work correctly the three beams of a colour picture tube must impinge accurately on the phosphor-dot triads of appropriate colours over the whole range of deflection. There is a beam for each colour, red, green and blue, and

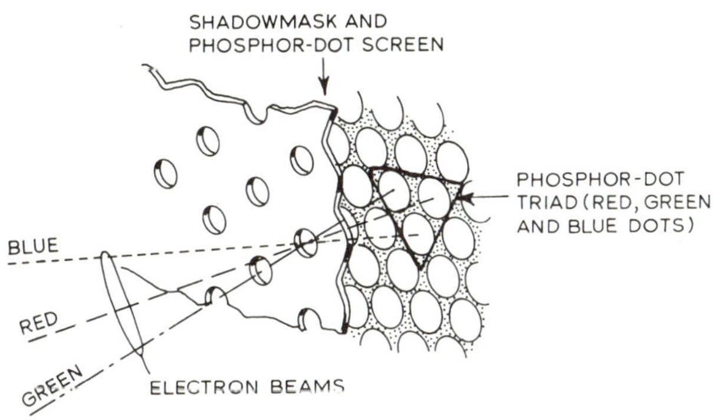

FIG. 10.1. *The principle of convergence.*

the red beam must activate only the red dots, the green beam the green dots and the blue beam the blue dots. Neither the beams nor the phosphor dots are coloured of course, the colours merely refer to the light emission for which they are responsible. To facilitate this action the majority of tubes in current use employ a shadowmask close to the inside of the screen which carries a large number of small holes corresponding to the screen phosphor-dot triads. The three electron beams pass through these holes as they are deflected in the usual way and then spread out slightly so that the dots of the correct triad are hit, as shown in Fig. 10.1.

CONVERGENCE

Although the shadowmask picture tube is accurately assembled, electronic and magnetic means are required to keep the three deflecting beams properly positioned for dot registration over the entire screen area. This is called convergence. Convergence in the middle of the screen (e.g., when the three beams are arriving along the axis of the tube neck) is handled by a set of permanent magnets or electromagnets carrying d.c., one corresponding to each beam. The field strength of each one is adjustable so that each beam can be 'bent' slightly if necessary to yield the condition shown in Fig. 10.1. This is called *static convergence* because the beams, when directed to the middle of the screen, can be regarded as undeflected—e.g., 'static'.

Convergence away from the middle of the screen is handled by a set of electromagnets energised with specially tailored current from the field and line timebases. Again, each beam is subjected to its own magnetic field, but the fields produced in this way change in strength as the beams are deflected away from the middle of the screen vertically and horizontally by an amount which is geared to the difference between the radius of beam deflection and the radius of the shadowmask. This is called *dynamic convergence*, and when the preset adjustments associated with the strength and nature of the fields are optimised the conditions as shown in Fig. 10.1 are substantially maintained over the whole screen area.

The static and dynamic convergence fields are effectively combined and are concentrated round each beam separately by pole pieces from the convergence unit (placed on the tube neck—see Fig. 10.2) which extend within the tube neck,

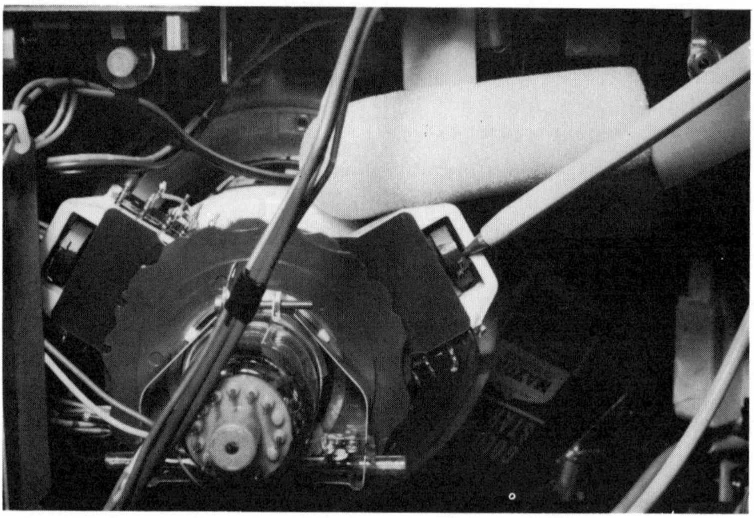

FIG. 10.2. *Convergence assembly in the BRC 3000 series signal-standard chassis. The pencil is pointing to one of the static convergence magnets (see text).*

and for dynamic convergence there is a coil or electromagnet associated with both vertical and horizontal scanning currents for each beam, making three pairs of coils in all. The preset dynamic convergence adjustments regulate both the amplitude and tilt of the current waveforms, and there can be up to a dozen such presets in single-standard receivers—more in dual-standard models.

PURITY

While the convergence fields correct the 'angle' of each beam separately for correct registration of the individual red, green and blue rasters, another entirely different magnetic field also across the tube neck but behind the convergence unit (e.g., closer to the tube guns) is used to adjust the originating angles of the three beams *together* so that they 'see' the appropriate colour dots after passing through the holes in the shadowmask.

The magnetic field for this is produced by a pair of ring magnets which can be rotated one relative to the other or both together on the tube neck, actions which regulate the effective field strength through the tube neck and the radial direction of the field to secure the correct positioning of the deflection centre along the tube axis.

The adjustment for this is called *purity* and the assembly concerned the purity magnet. The term purity is used since when the adjustment is correct the raster will appear as pure red (without contamination from the other two colours) when the red gun only is running (e.g., when the blue and green guns are switched off), and likewise with either one of the two other colours. When the three guns are running and correctly adjusted the raster will appear as white on a pure tube, but contaminated with spurious colours when impure.

Overall purity is affected by the positioning of the scanning coils on the tube neck, and it is common practice to slide the coils a little along the neck towards the tube base and then adjust for purity on a red raster until a patch of pure red appears right in the middle of the screen (possibly within a 'sea' of spurious colours which should be ignored). When the correct purity adjustment has been established the patch of red will increase in area as the scanning coils are slid back towards the tube flare and eventually fill the screen with a pure red raster.

Fig. 10.3 shows the purity magnets of a Decca single-standard colour set, while Fig. 10.4 shows a Mullard scanning assembly. The wing-nuts on this allow the coils to be slid along the neck for purity adjustments. Similar wing-nuts can also be seen in Fig. 10.3.

FIG. 10.3. *The pencil in this picture is pointing to the purity rings in the Decca single-standard receiver.*

FIG. 10.4. *Mullard scanning coil assembly, showing the adjusting wing-nuts (see also Fig. 10.3).*

It will be appreciated, of course, that correct registration occurs only when the convergence is optimised. Under this condition the three beams converge at a common point on the shadowmask and illuminate the three phosphor dots of a triad (in practice a group of about three triads is illuminated), this giving the correct registration of the three colours. Misconvergence means beam misalignment. If, for example, the blue beam is misconverged, the red and green beams would be correctly striking the red and green phosphors of a common triad (which would give a yellow display), while the blue beam would be illuminating the blue dot of a *different* triad, thereby yielding a *displaced* blue display. This is misregistration. Correct registration thus occurs when the three beams illuminate the phosphor dots of *common* triads, thereby yielding a white display when the beam currents are adjusted accordingly. It will be noted, therefore, that the purity remains substantially good even under conditions of mild misconvergence.

We shall see in a minute that the crosshatch and dot generator is necessary for convergence adjustments, but before seeing why this is so let us have a look at grey-scale tracking.

GREY-SCALE TRACKING

A shadowmask picture tube can be regarded as three tubes in one, each one producing a primary colour (e.g., red, green and blue) display. For a monochrome display (e.g., picture 'white') the three tubes have to be operating simultaneously and the beam currents must be proportioned relative to the efficiency of the different colour phosphor dots to yield the required 'colour temperature', nowadays called Illuminant D, corresponding approximately to the north sky at noon and to the white light produced when a solid object is heated to a temperature of about 6,500°C.

Colours are produced from the three primaries and through their complementaries (e.g., yellow complementary of blue, cyan complementary of red and magenta complementary of green) by changes in the current of the three beams and hence by changes in the intensity of dot illumination. Obviously, the display would be in red if only the red gun was running. The same with the other two colours. White, we have seen, is produced when all three guns are running; and if one of the primary colours is removed from a white display we get a display in the complementary colour of the hue removed. Thus we get a yellow display when blue is removed from white. The saturation of a colour is reduced when white light is added to it. This means, then, that blue would be fully saturated with only the blue gun running and progressively desaturated as the other two guns are brought into action and their beam currents increased. We would thus eventually get back to white.

The luminance or 'brightness' of a display is related to a signal composed of 30% red signal, 59% green signal and 11% blue signal. If the amplitude of a signal made up of these proportions is reduced, the display in monochrome (the luminance signal carries all the monochrome information) would fall in brightness, the same as when the contrast control of a monochrome set is turned down. Note, though, that the proportions must be maintained over the range of control to avoid the monochrome display changing towards a colour.

The same applies to the brightness control proper. This must adjust the beam currents of the three guns very closely in step to avoid the red, green and blue intensities changing from those required for a monochrome display to those for a colour display. This is grey-scale tracking.

Let us suppose that we adjusted the beam currents at a specific brightness level to give a white raster and found that as we reduced the brightness setting the raster veered from white, say, towards red. This would imply that the intensity of illumination of the blue and green phosphor dots is falling behind that of the red dot with increasing brightness.

Best technicians adjust the white temperature by means of a comparative source of illumination corresponding to Illuminant D, the correction required to delete an unwanted tint then being made according to the following table.

Display tinted	*Correct by*
Red	Increasing blue and green
Blue	Decreasing blue
Green	Decreasing green
Magenta	Increasing green
Cyan	Decreasing blue and green
Yellow	Increasing blue

However, when such adjustments have been carried out the display could veer from the correct colour or from white on a signal or when the brightness control is adjusted when the three guns fail to track over the full dynamic range of the picture (e.g., from black to peak white). This is where the grey-scale tracking comes in, and the crosshatch and dot generator commonly provides a white raster for making such adjustments.

Most colour receivers have two adjustments for the blue and green guns and one for the red one to optimise the grey-scale tracking. The blue and green guns each have a preset at the high-brightness end of the range and at the low-brightness end. The first adjusts the video drive at the tube cathode and the second the potential at the first anode, the plan being to get the characteristics of the three guns to match as accurately as possible over the dynamic range. The red gun only has a first anode preset, the cathode generally being connected direct to the video (luminance) amplifier, since the red phosphor is the least sensitive.

FIELD INSTRUMENTS

All the adjustments so far described might well have to be undertaken at the home of a viewer, so the instruments of the first part mentioned at the start of this chapter must certainly include a crosshatch and dot generator. Minor adjustments to convergence can, it is true, be made by using the transmitted Test Card F, but for first class results a crosshatch (if not with 'dots') generator is essential. I have already given some information on this kind of instrument (page 82), but to recapitulate, fairly widely spaced dots are useful for adjusting the static convergence, while a crosshatch is more suitable for dynamic convergence adjustments.

It is obviously impossible to detail the various adjustments of the different types of colour sets, but set makers are now getting this down to a fine art with well-designed plans showing the various presets to be adjusted and the effects of the adjustments on the crosshatch display. The presets of some sets are themselves well labelled, it being necessary merely to follow arrowed lines from control to control, adjusting each in turn for the indicated result on the crosshatch display.

When a colour set is installed the tube may be seen to be impure. A typical symptom being a patch of the wrong colour towards one corner. This can be cleared only by adjusting for purity, but it is possible that spurious magnetic fields are affecting the beams, so before the procedure for purity is put in hand the tube and surrounding metal items in the room should be demagnetised—called degaussing. All modern colour sets are equipped with an inbuilt degaussing coil located round the tube flare which is energised with fairly high power pure a.c. (from the mains supply) each time the set is switched on and, with some sets, when the channels are changed. The a.c. gradually falls from the peak switch-on value and after a short while diminishes virtually to zero and with it, of course, the degaussing field. A collapsing a.c. magnetic field is necessary to extract residual magnetism.

However, the inbuilt degaussing arrangement is rarely sufficiently strong to pull out strong residual magnetism and is certainly no good for degaussing metal items near the receiver. For these activities an external degausser is required. This, then, is another field instrument. Various types are available designed for working direct from the mains supply. Owing to the high current in the coil of which they are composed they must only be used intermittently to avoid overheating. Degaussing coils are normally required for only a couple of minutes' treatment round the front of the picture tube to degauss the shadow-mask. The tube face is scanned with the field from the instrument and then taken several feet away before it is switched off. On no account should the tube neck assemblies come under the degaussing field influence, but it is not necessary to switch the colour set off while the tube is being degaussed. The resulting colour displays are very impressive to the onlooking viewer! Similar treatment can be given to nearby metal objects, but make sure there are no tape recordings in close proximity, and it is no good degaussing a radio set to kill the field from the loudspeaker! If the tube is still impure, then adjustments for purity will have to be made.

It would seem that colour sets of early vintage were more prone to external magnetic fields than are today's models. Indeed, just changing the position of the set relative to the Earth's magnetic field called for a degaussing operation, but rarely is this necessary these days. The author can run his colour set anywhere in the room without disturbing the purity.

For adjusting the purity and grey-scale tracking the crosshatch generator should desirably yield a signal for producing a white raster. Of course, the aerial can be removed from the set and the brightness turned up to get a non-modulated display, but this is often too noisy (e.g., containing grain) to make the best adjustments.

Another useful field service instrument is the e.h.t. voltmeter, but like the oscilloscope this is needed more by the service technician than the installation technician. However, it is sometimes necessary for the installation technician to adjust the white colour temperature to the viewers' tastes and for this an Illuminant D comparator can save a great deal of discussion!

To summarise, therefore, the essential instruments for installation and initial setting-up are:

1. *Crosshatch/dot generator.* This must be dual-standard if it is likely that the sales programme includes sufficient number of dual-standard models to justify its purchase. Actually, by the time this book is in print the vast majority of models sold will be single-standard, so technicians and dealers just getting involved with colour receivers would do best to purchase one of the more recent single-standard generators. These are less costly and less complicated (with less to go wrong, as with the sets themselves) than the earlier dual-standard versions. For field activities a transistor battery-powered model is significantly more useful than a mains-powered counterpart. I have already mentioned the importance of good locking ability, and if possible it is just as well to try a model out before making the pur-

chase to ensure that it conforms. To ease grey-scale tracking adjustments, a grey-scale pattern facility is a very useful addition.

2. *External degausser.* At the time of writing a number of models of this instrument readily available when colour was first launched have been withdrawn. This seems rather amazing for the author's experience has shown that degaussing is still as necessary now as it was at the commencement of the colour service, in spite of the fact that recent sets are equipped with internal degaussing. However, if the worst comes to the worst, one can always construct a degausser from an old speaker field winding or similar large inductor. A bulk eraser for tapes could also be used at a pinch.
3. *Illuminant D comparator.* Different viewers seemingly see white light differently, and having a source of definite white light can save a lot of time and argument. It only needs a very slight movement of the red, green or blue preset to introduce a vague tint of primary or complementary hue.

The field service technician will require instruments on top of those described above, but since the extra ones are contained within the second part mentioned at the start of this chapter they will be covered under workshop instruments which we shall now look at.

WORKSHOP INSTRUMENTS

The workshop will also require an instrument capable of producing crosshatch and dot patterns, and where the cost can be justified this should possess colour encoding facilities, giving colour bars or a rainbow display. Again, this type of instrument has already been mentioned.

From time to time complete colour set realignment will be necessary, and for this critical operation only visual alignment can be considered. Thus a suitable wideband wobbulator with oscilloscope are essential instruments. The wobbulator should sweep easily over the channel bandwidth and the sweep should be activated either by the timebase of the oscilloscope or from 50 Hz mains (60 Hz in America) which also provides the X sweep of the oscilloscope, via a suitable phasing network.

The oscilloscope should have a Y bandwidth of at least 6 MHz to the 3 dB points so that the full video signal can be displayed without attenuation of the colour subchannel components and higher video harmonics. The sweep should also suit the Y bandwidth, going down to at least 1 μs or 0·1 μs with X expansion. Repetitive and trigger sync facilities should be incorporated and also locking to television field signals.

When the instrument is to be used for field servicing portability is a key factor, and it is much better if the instrument can be made active without having to connect it to the mains supply. A good portable battery/mains instrument thus has much to be said for it.

In the workshop (and sometimes in the field) it is necessary to measure the e.h.t. potential on the final anodes of the shadowmask tube. This can go almost up to 30 kV so the instrument must feature a very sensitive (low current) meter movement of at least 20,000 ohms-per-volt. Instruments such as the Avo Model 8 and the SEI/GEC Selectest Model Super 50 are of this sensitivity and can thus be used with a high voltage probe. Such a probe designed for the Selectest is shown in Fig. 1.15, while Fig. 10.5 shows the Selectest Super 50, which is a multirange meter.

The measurement of e.h.t. voltage demands extreme caution since the potential is nominally 25 kV at a power of 30 W. This is potentially lethal, so one should never be tempted to draw sparks from the e.h.t. circuits of colour sets. Moreover the capacitance of the picture tube can store sufficient energy to be a killer, so the capacitance (afforded by the inner and outer conductive

FIG. 10.5. *Selectest multirange instrument with which can be used a high voltage probe.*

coatings of the picture tube) must always be discharged before working in the set, and it is noteworthy that the capacitance can retain the charge for a very long period after the set has been switched off.

Discharge the tube (with the set off, of course!) with the metal blade of a screw-driver having a very well insulated handle by connecting an earthing link between the blade and the metal chassis of the set and then pushing the blade underneath the disc of the e.h.t. connector. It is highly desirable to keep the capacitance under shorting conditions by preparing a shorting link with a crocodile clip one end and an e.h.t. tube connector the other end. With the screwdriver holding the short while the e.h.t. connector is pulled off, the e.h.t. connector of the previously prepared shorting link should then be pushed on to the tube final anode connector with the crocodile clipped end well terminated to the metal chassis. This will reduce the risks to a minimum, but care must be taken while making the original short and connecting the shorting link not to make inadvertent connection with the final anode circuit.

The workshop will also require its own degaussing device and, of course, all the usual monochrome test instruments, possibly including a signal generator for putting a marker pip on the displayed response curve during alignment. Other useful instruments include an Illuminant D comparator and an instrument for observing the illuminated phosphors on the tube screen so see that the beams are landing correctly. Such an instrument is the Philips Beam Landing Pattern Scope (called BLP Scope), illustrated in Fig. 10.6. It is basically a periscope and microscope combined, allowing a technician to view a small portion of the screen from the rear of the set while adjustments to convergence, etc. are

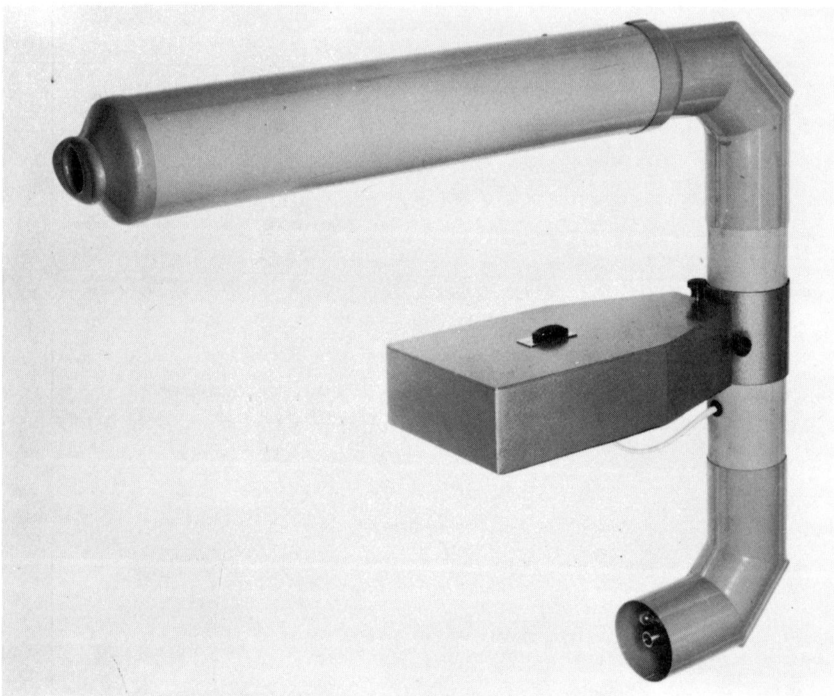

FIG. 10.6. *Philips Beam Landing Pattern Scope (see text).*

being made. Magnification is thirty times, which means that the screen phosphors become visible. The eyepiece is located at one end of a long tube, while the end of a smaller tube at right-angles carries an objective lens and small bulb. The case clamped at the centre of the shorter tube houses the battery for lighting the bulb and an on/off switch at the top. This usually rests on the top of the receiver cabinet while the adjustments are being made.

INSTRUMENTS FOR THE FIELD SERVICE TECHNICIAN

We can now see what instruments in addition to those detailed under the section dealing with field instruments are required by the service technician, as distinct from the installation technician. Probably the most important is the oscilloscope which, as already mentioned, is best battery/mains powered. The field operation will be essentially that of fault diagnosis, the actual repair probably resolving merely to the fitting of a replacement 'module'.

The service manual details the various waveforms to be expected at the test points of a correctly operating receiver, so a quick check round with an oscilloscope will soon bring to light abnormal working. Indeed, the oscilloscope is just as important for colour receiver servicing as the multimeter is for radio receiver servicing. The Y axis should be in reasonable calibration, since a fault might be located by a signal of insufficient or excessive amplitude, and the sweep speed should be indicated in terms of time (s, ms or μs) per centimetre of horizontal sweep.

For some applications a low capacitance test probe will be necessary to prevent the reactive loading of the oscilloscope from affecting the circuits under test (which can arise, for example, from the capacitance of a screened connecting cable). A low capacitance probe isolates such reactive influences from the test circuit while still passing the signal information to the Y input of the oscilloscope. However, some attenuation in the probe is inevitable, and this has to be taken into account when measuring test signal amplitude.

We shall need a pattern generator, of course, and while a simple and inexpensive crosshatch/dot generator is suitable for carrying out installation work, a colour-encoded instrument might be necessary to produce signal for oscilloscope analysis. An instrument producing the standard colour bars can be very

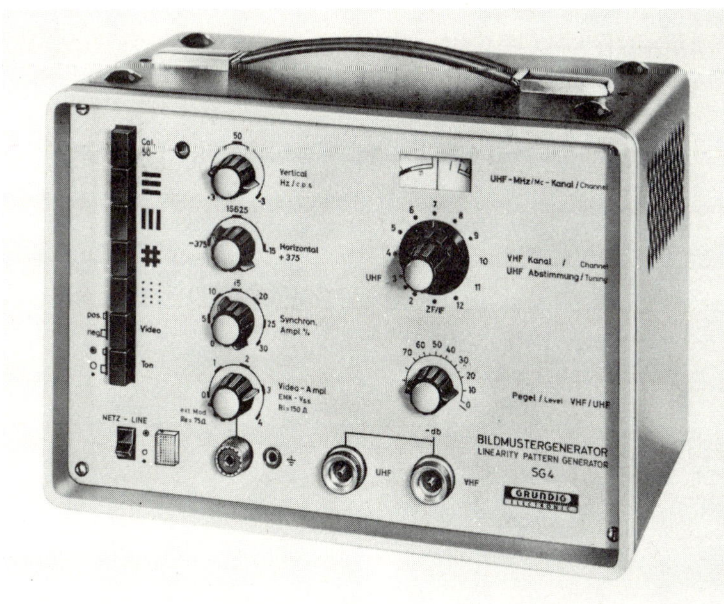

FIG. 10.7. *Grundig pattern generator. The type of patterns that this instrument can produce can be seen by the side of the selector push-buttons at the left.*

useful in this context since some fault conditions can also be appraised merely by viewing the colour bars on the screen of the receiver without any other instruments (e.g., see my book *Colour Television Servicing*, by the same publisher). Such a generator produces signals which very closely simulate those transmitted by the BBC and ITV; but of course the cost is higher than for a simple non-colour pattern generator.

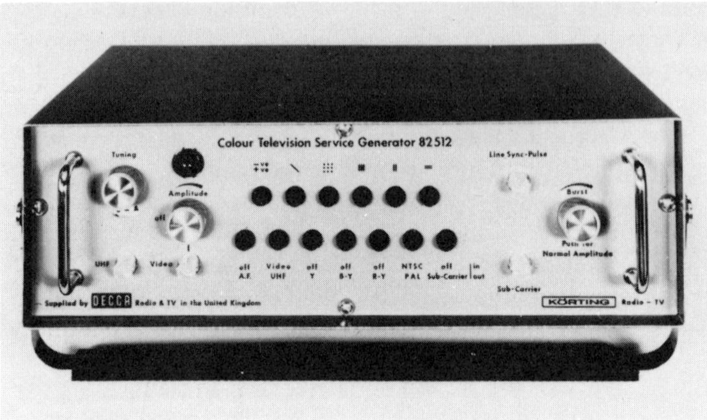

FIG. 10.8. *Fully colour-encoded pattern generator by Körting, now model number 82514 (UK distribution by Decca). This provides signals at both NTSC and PAL standards.*

Fig. 10.7 shows a Grundig pattern generator, while Fig. 10.8 shows a fully colour-encoded generator by the German Körting firm (distributed in the UK by Decca Special Products). Samples of the colour signals produced by the Körting Model 82512, now Model 82514 (see also Fig. 5.6) are shown in Fig. 10.9. That at (*a*) is a colour burst on the porch to the line sync pulse and that at (*b*) one and half lines of signal due to the standard colour bars, also showing the

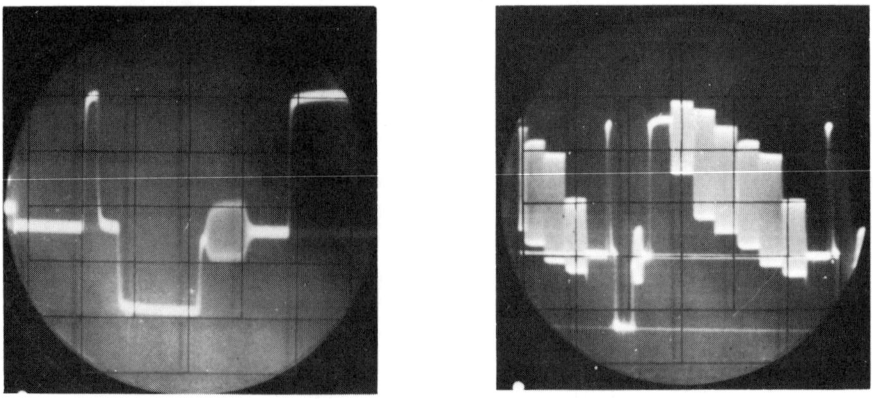

(a) (b)

FIG. 10.9. (a) *Line sync pulse and colour burst and* (b) *one and half lines of colour bar signal, showing one line sync pulse and burst (see also Fig. 5.6).*

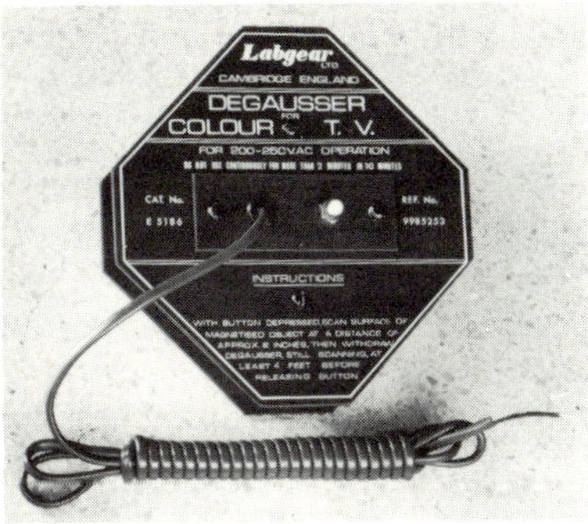

FIG. 10.10. *Example external degausser. This one is supplied by Labgear Ltd.*

FIG. 10.11. *Oscilloscope by Grundig Electronics.*

line sync pulse and colour burst. It will be appreciated that these signals closely resemble those from the transmitting authorities.

A sample degausser is shown in Fig. 10.10. This one is made by Labgear Ltd and the press-button controls the mains power. Also from the extensive range of Grundig equipment is the type G8/13 oscilloscope shown in Fig. 10.11.

To conclude this survey, mention again must be made of the signal strength meter which is another important instrument for the field service technician, bearing in mind that quite a few colour television complaints are due to nothing more than insufficient aerial signal. The signal strength meter quickly proves this point.

CHAPTER 11

Instruments for Audio

SINCE the incredible rise in the popularity of hi-fi and good quality audio systems, no doubt encouraged to some extent by the more sophisticated use of solid-state devices, the development of new such devices specifically for audio applications and possibly the public wanting a change from television, radio dealers and their technicians have had increasing call to test and service audio equipment. Indeed, it would appear that a new 'testing science' is growing in the audio world, and the need to secure a basis from which to work is being revealed by the many new audio specifications which are being evolved.

This chapter surveys the various instruments required for this new-generation audio testing, while giving information on some of the major tests. But firstly what instruments are required?

AUDIO INSTRUMENTS

Most audio testing is dynamic which means that the tests are made relative to signals in the system. A very important instrument, therefore, is that which generates audio signal—the audio oscillator or generator. A fair amount of information on the design of this kind of instrument has already been given in Chapter 4 where it was intimated that a chief parameter is a very low distortion sine wave adjustable in frequency from at least 20 Hz to 20 kHz. However, since quite a few of the better quality hi-fi amplifiers have frequency and power responses extending well above 20 kHz it is desirable for the audio generator to yield low distortion signals at least to 30 kHz.

Music waveforms, though, are considerably different from pure sine wave so the generator should desirably produce a signal which gives a better simulation to music waves, and the least complex in this respect is the square wave. A music wave can be analysed into a multiplicity of sine wave components and the characteristic waveshape of a music signal arises from the integration of the sine wave multiplicity in specific amplitude and phase relationships. A square wave is similarly composed of a multiplicity of sine waves of odd-harmonic relationship to the fundamental and at diminishing amplitude, as shown in Fig. 11.1. Here sine-wave components up to the seventh harmonic only are shown; and it is noteworthy that the quality of the resultant square wave is

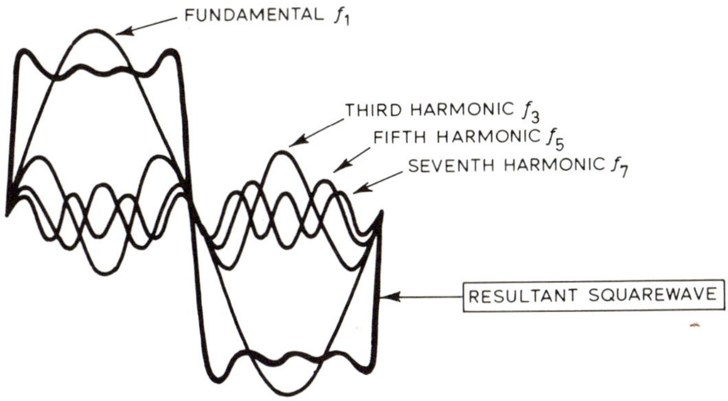

FIG. 11.1. *Showing the composition of a square wave. The greater the number of odd harmonics, the better the square wave shape.*

progressively improved as more odd-numbered harmonics are added in the manner

$$Af_1 + \frac{A}{3}f_3 + \frac{A}{5}f_5 + \frac{A}{7}f_7 + \text{etc.}$$

where A is the amplitude of the waves. The nature of the phasing of the sine wave multiplicity is shown in Fig. 11.1.

Clearly, then, if a square wave is handled by an audio system without significant form distortion the system must be passing the component sine waves without changing their amplitude or phase. Moreover, the rapidly rising and falling leading and trailing sides of a square wave subject the amplifier, for example, to rapidly changing currents which, in some designs, can precipitate overshoot and ringing effects, described later. The square wave form is preserved provided the sine wave harmonics up to about the tenth are passed by the system without significant amplitude or phase change. Thus a system which passes, say, a 10 kHz square wave could be assumed to have a frequency and phase-linear bandwidth up to about 100 kHz. Usually, however, even the best amplifiers exhibit some corner rounding of a square wave passed through at a fundamental of 10 kHz. This is due to attenuation of the higher-order harmonics, since there is generally little need for an audio amplifier to be wide open towards full (or half) power at 100 kHz!

Obviously, a music waveform is not of a square wave naturė. However, as shown in Fig. 11.2 it does carry some of the characteristics of a square wave, especially in terms of fast rising and falling amplitudes, resulting in transients.

The sine wave generator, therefore, should have facilities for producing square waves at fundamental frequencies equal to those of the sine waves. As we saw in Chapter 6, the rise time of an amplifier is an important consideration, and this is applicable, too, to the amplifier and square wave producing circuits of the audio generator or oscillator, which means a wide passband.

The audio generator must be complemented with an instrument capable of 'reading' the signal in or at the output of the equipment being tested, and one such instrument is the audio millivoltmeter (see Chapter 4). Another which not only 'reads' but also displays the signal is, of course, the oscilloscope. These two instruments, therefore, are equally as important for testing in audio equipment as the sine/square generator.

We often require to measure the r.m.s. voltage of a sine wave across the output load of an audio amplifier on the one hand and the low-level signal from, say, a gramophone pickup on the other. Thus the audio millivoltmeter should read at

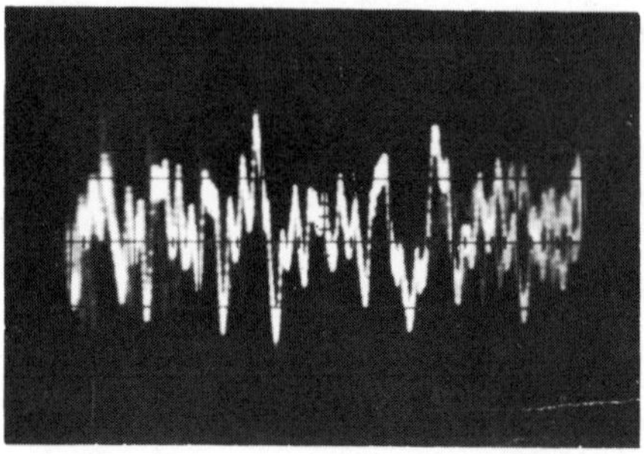

FIG. 11.2. *Oscillogram of a music wave.*

least up to 20 V r.m.s. (corresponding to 25 W into 15 ohms) and down to 1 mV r.m.s. These are full-scale readings. In amplifier testing it is also common practice to read noise signals on a r.m.s. basis across a load over the amplifier's inherent bandwidth, so a low-reading millivoltmeter is essential for this application. In low-level amplifiers such noise signals can only be measured with a highly sensitive millivoltmeter reading down to, say, 10 μV full-scale; but tests like this are few and far between, so a good instrument with a f.s.d. down to 1 mV r.m.s. is generally adequate for the majority of applications.

One big problem when measuring very low-level signals is induced or coupled hum signals from the mains supply or power components of the equipment under test and care must be taken by suitable screening, etc. to keep this at a minimum. Hum-loop conditions can also develop when the audio millivoltmeter is mains powered and when it is connected to a mains-powered amplifier. This can be avoided by careful earthing at one point only and by the use of battery-powered test equipment. I prefer battery-powered audio test equipment in view of its portability and freedom from hum-loop conditions. However, battery costs can be high when the equipment is in protracted operation and for general workshop use mains-powered equipment has a great deal to commend it, and is perfectly adequate when used correctly.

The oscilloscope should have the features outlined in Chapters 6 and 7. The Y bandwidth should yield a rise-time parameter at least equal to that of the square wave generator, which commonly means a bandwidth of at least 3 MHz to the −3 dB points. An oscilloscope selected for colour television servicing is generally ideal for audio testing and servicing applications.

The X sweep should run to at least 1 μs/cm (corresponding to a 1 MHz sine-wave display within 1 cm of sweep) and X expansion allows this to be increased sometimes up to ten times (e.g., 0·1 μs). The triggering and repetitive sync should

FIG. 11.3. *Avometer Model EA113. This solid-state electronic multimeter is finding application in the audio testing laboratory and workshop since it reads audio signals over a range of 10 mV to 1,000 V. It also measures audio current up to 10 mV and a.c. up to 3 A f.s.d. over a number of ranges. Other measurements are d.c. voltage, d.c. current and resistance. It operates from four ZM9 Mallory mercury cells.*

be solid and consistent at the highest frequency and since it is sometimes desirable to photograph audio waveforms there should be provision for the fitting of an oscilloscope camera.

Y channel sensitivity should cover from 100 V or so per cm down to 100 mV per cm or less over a number of calibrated ranges. To check the Y attenuator accuracy some instruments deliver a sine wave calibration voltage at a facia terminal.

An example of a dual-trace portable oscilloscope suitable for servicing or laboratory use is the Advance OS 1000, shown in Fig. 11.4. The display is from a 10 × 6 cm rectangular tube running at 4 kV on the final anode, available with standard or long persistence phosphors. Calibration is by means of the usual graticule.

Y1 and Y2 (vertical) amplifiers give separate control of the two traces, with a sensitivity of 5 mV/cm to 20 V/cm, in 1:2:5 sequence switched and 2·5:1 fine control. Y bandwidth is d.c. to 15 MHz (−3 dB) or from 2 Hz with a.c. input coupling, there being the usual choice of a.c. and d.c. Rise time works out to 23 ns. Maximum Y input is ±400 V peak. An interesting feature is a Y2 output which allows a Y1 and Y2 cascading, resulting in a maximum sensitivity of

FIG. 11.4. *An OS 1000 oscilloscope by Advance Electronics.*

1 mV/cm over 5 Hz to 5 MHz, single channel.

Display is alternate sweep or chopped mode (250 kHz), the former adjustable from 0·5 µs/cm to 0·5 ms/cm, and the latter from 1 ms/cm to 1 s/cm. X–Y mode operation is also possible via the Y1 and Y2 inputs over a bandwidth of d.c. to 500 kHz.

X deflection timebase goes from 1 s/cm to 0·5 µs/cm in 1:2:5 sequence switched, with a fine control over a range of 2·5:1, giving greater than 25 sec slowest sweep. An X gain control provides expansion of 10 times, corresponding to a maximum sweep rate of 50 ns/cm.

Trigger modes are internal Y1 + or −, internal Y2 + or −, external + or − and line + or −, and couplings are provided for a.c., a.c. fast and television field, the latter making it possible, for example, to view the colour bar lines at the top of Test Card F (see, for example, my Newnes-Butterworths book entitled *Colour Television Servicing*).

Facilities are also provided for external Z modulation (to the display tube grid), manual and auto level control, with bright line free run in absence of signal, adjustable internal sensitivity, external X input, external trigger, calibration output (square wave of 1 V p–p with about 30 μs rise time), probe set (timebase gate waveform of + 10 V) and ramp (+ 8 V from about 6·8 k at rear panel).

This is the kind of instrument that is particularly suitable for colour television servicing and for serious work on hi-fi audio systems. Powering is from the mains over a range of voltages from 95 to 260 and from 45 to 440 Hz.

The next most important instrument—indeed, just as important as the others just detailed if serious audio testing is to be undertaken—is the distortion test set. This can be a complete instrument on its own or it can be merely a tuned filter arrangement which notches out the fundamental sine wave leaving the distortion components which are measured and analysed by a millivoltmeter and oscilloscope. Some distortion test sets also include not only the readout instrument but also the sine wave generator tunable over the audio spectrum. Such instruments read essentially total harmonic distortion (THD), automatically summing the power in the harmonic components remaining after the fundamental has been deleted.

A more sophisticated instrument is the wave analyser which selects just one of the harmonics for measurement and comparison in strength with the fundamental frequency. Information has already been given with regard to the various types of distortion measurement in Chapter 1, so there is no point in repeating them here.

Most service technicians are content to measure the THD to ensure whether an amplifier is up to specification, and THD is a fair way of comparing the distortion performance of one amplifier against another. Intermodulation distortion (IMD) is another useful measurement also detailed in Chapter 1, and it is often useful to run this sort of test in parallel with that for THD.

Clearly, the distortion in the source signal must be significantly below that likely to be introduced on to the signal by its passage through the test amplifier,

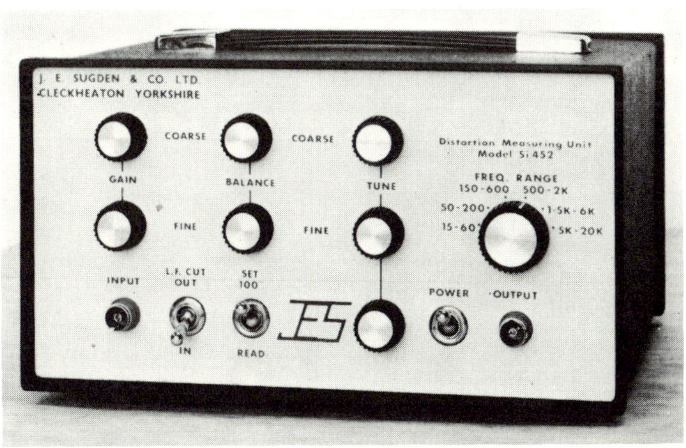

FIG. 11.5. *Model Si 452 Distortion Measuring Unit by J. E. Sugden.*

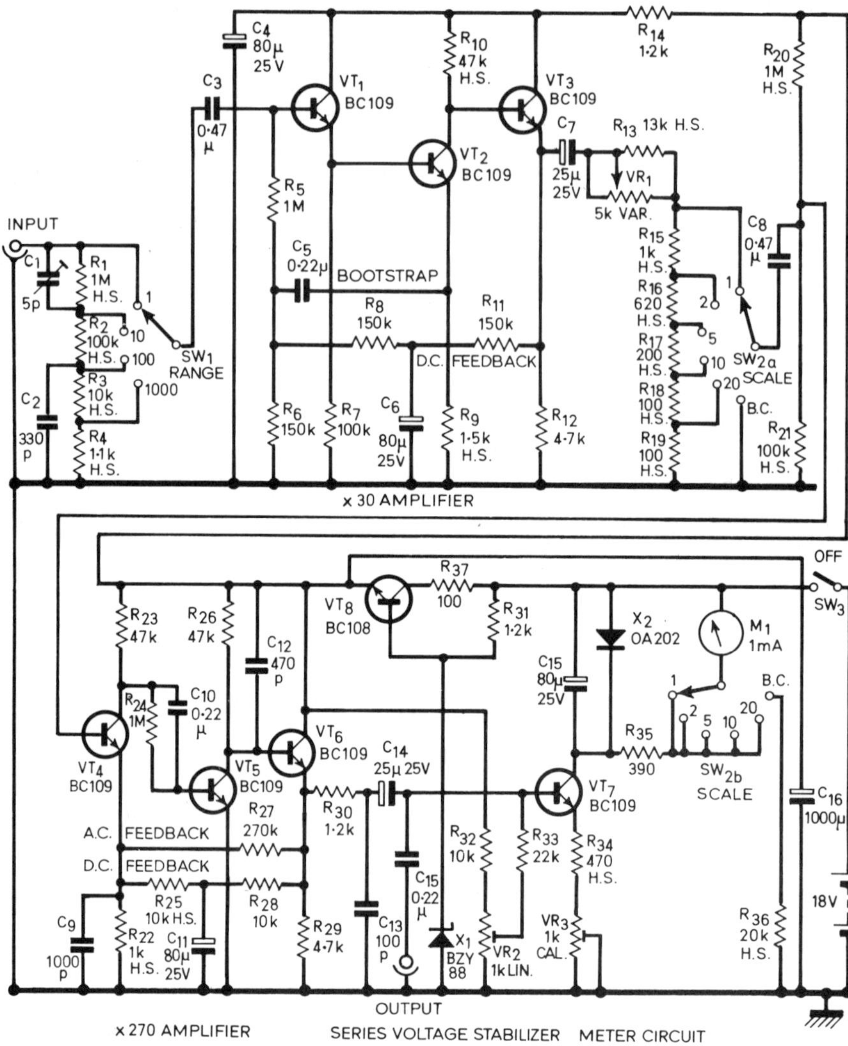

FIG. 11.6. *Circuit diagram of the Distortion Measuring Unit shown in Fig. 11.5. (see text).*

for example. The British Standards Institution implies that the test-tone distortion should be at least 1/5th that of the amplifier distortion. To measure THD down to 0·05 %, therefore, the sine wave generator should yield no more than 0·01 % distortion. This, then, reveals the need for a very low distortion generator, but it is possible to employ a generator of higher intrinsic distortion and filter the signal prior to its application to the equipment under test. More will be said about this later.

The J. E. Sugden distortion measuring unit is depicted in Fig. 11.5. This is the third of the trio of battery-powered audio instruments made by this firm (the audio millivoltmeter and sine/square generator have already been explored). For the sake of completeness as well as interest, the circuit of the instrument is given in Fig. 11.6. This is based on an *RC* notch rejector in a feedback loop to sharpen the edges of the characteristic. The filter exhibits a very high *Q* providing rejection greater than 80 dB. The input circuit also contains a switchable two-stage high-pass filter for reducing the low-frequency noise and hum from the readout. The signal amplitude can be adjusted to suit the circuit by the input attenuator, the optimum sensitivity being 1 V r.m.s. The high notch filter attenuation allows THD measurements down to 0·01 % under optimum conditions (0·05 % 'nominal'). Notch filter tuning range is from 15 Hz to 20 kHz and, like the other instruments of the range, the distortion measuring unit Model Si 452 works from three internal 4·5 V batteries, such as Every Ready AD28.

Various other firms are producing test equipment to meet the demands of audio measurements, including the hi-fi pioneer Arthur Radford from his Radford Laboratory Instruments Limited firm at Bristol. Two notable Radford instruments are the low distortion oscillator and the distortion measuring set.

(a)

(b)

(c)

FIG. 11.7. *Audio test instruments from the Rogers range.* (a) *Low distortion oscillator, Type S324.* (b) *A.F. millivoltmeter, Type AM342.* (c) *Distortion factor meter, Type DM344A. (Rogers Developments (Electronics) Ltd.)*

The first yields pure sine wave signals with the remarkably low distortion of 0·005% from 200 Hz to 20 kHz, rising only to 0·015% from 10 Hz to 100 kHz. Frequency coverage overall is from 5 Hz to 500 kHz in five ranges. The instrument also provides square wave signals with a rise time less than 0·1 μs. Design is hybrid, using both valves and semiconductors.

The distortion measuring set gives THD readout and is self-contained. It is capable of recording distortion products as low as 0·002%, readings being direct from a calibrated meter scale.

Rogers Developments (Electronics) Limited, another hi-fi manufacturer, have also recently introduced a range of audio test equipment for the laboratory and service technician. Some of the units in the range are interrelated, thereby enabling the build-up of complete test systems on a modular principle.

Three instruments are a distortion factor meter which will measure from 100% THD down to 0·1% at full-scale deflection in seven switched ranges, a low distortion oscillator with a typical 1 kHz THD of 0·015% and an a.f. millivoltmeter with full-scale readouts from 1 mV to 300 V in twelve ranges.

These instruments, pictured in Fig. 11.7, are battery powered, though mains power units, operating from 100–125 V and 200–250 V, 50/60 Hz, are available as optional extras.

With the rising interest in high quality audio reproduction there is a growing demand for equipment of this kind, and portability can sometimes be a key factor. Battery powering is also useful under certain test conditions for reducing the tendency towards hum loop conductions, etc.

To summarise, therefore, the most important instruments for audio testing are:

1. sine/square generator.
2. audio millivoltmeter.
3. oscilloscope.
4. distortion measuring set.

With these instruments plus those already in the service workshop the vast majority of tests and adjustments can be made in contemporary audio equipment. Let us now consider some of the tests.

POWER MEASUREMENT

The setup for checking audio power output is shown in Fig. 11.8. The amplifier should have its volume control at maximum, tone controls 'flat', balance control centre and filters 'out' and the signal should be applied to a non-equalised input, such as that for 'radio' or 'auxiliary'. The output terminals of the channel under test should be terminated with a resistive load within the loading range of the amplifier and of a power rating to handle the full steady-state power without burning up. The r.m.s. voltage across the load should be measured on a suitable audio millivoltmeter, while at the same time the waveform should be monitored on an oscilloscope.

Making sure that the generator signal is turned right down, the amplifier

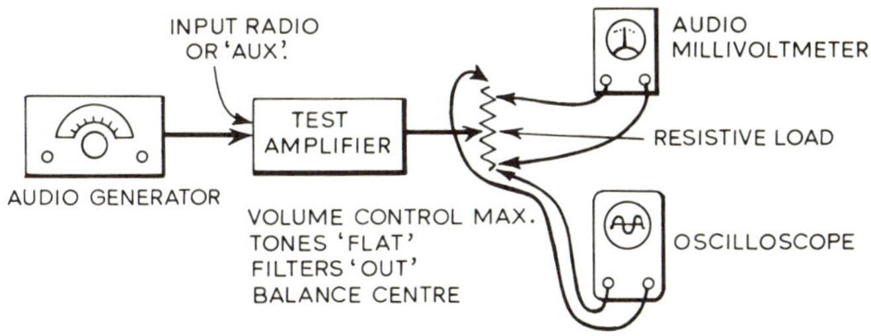

FIG. 11.8. *Typical audio test setup (see text).*

should be switched on and the generator signal gradually turned up so that the oscilloscope can be synchronised. The input should then be progressively increased and the Y attenuation of the oscilloscope and the millivoltmeter switching adjusted to cater for the increasing signal amplitude. Eventually the sine wave will be seen to clip at one or both peaks (Fig. 11.9) which indicates power amplifier overloading. The input should then be slowly turned down until the clipping *just* disappears and the pure sine wave appears once again (Fig. 11.10). At that output the r.m.s. voltage given by the millivoltmeter should be noted. The audio power in the load is then equal to the voltage squared divided by the load's resistance in ohms (e.g., $W = E^2/R$). This is the maximum steady-state or r.m.s.-based power in watts that the amplifier channel can deliver to the load value employed, commonly called the power capacity of the amplifier.

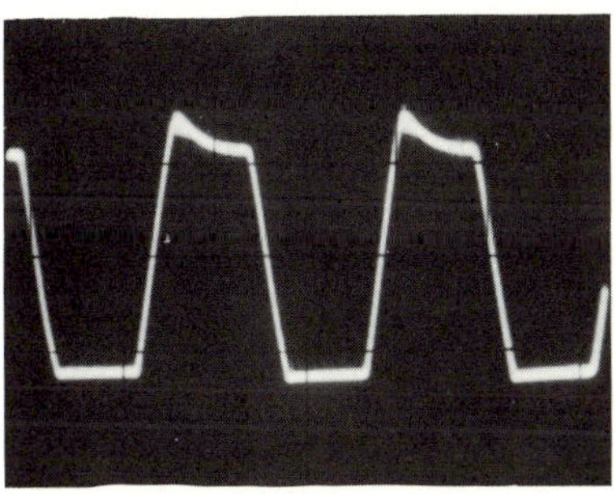

FIG. 11.9. *Sine wave clipping due to power amplifier overload.*

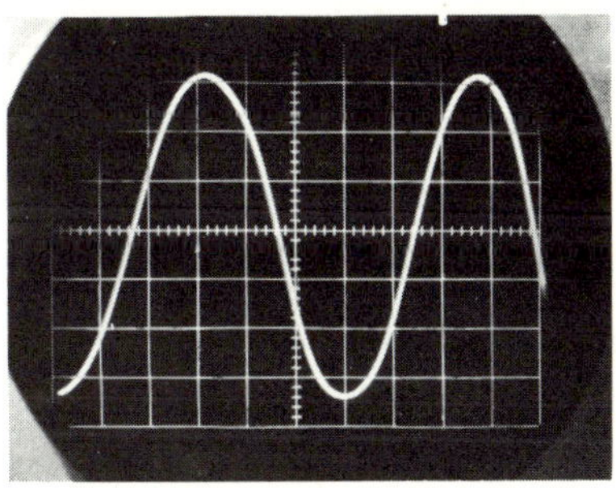

FIG. 11.10. *The clipped wave in Fig. 11.9 is restored to a sine wave as shown by turning down the input signal. Power capacity is measured just prior to the clipping level.*

It is the same sort of power as taken, for example, by an electric fire or bulb. Peak power is two times the steady-state power merely because the peak voltage of a sine wave is $\sqrt{2}$ times the r.m.s. voltage. Although peak power is sometimes stated, possibly to give the impression that an amplifier is more powerful than it really is, this mode of expression is deprecated.

Power capacity is commonly taken at 1 kHz and sometimes at the spectrum extremes, such as at 40 Hz and 10 kHz. Transistor amplifiers generally produce increasing power into reducing load values, so it may be worthwhile to check the power capacity at load values of 16, 8 and 4 ohms. It is sometimes desirable to measure the power capacity with both channels driven simultaneously. Normally, this reduces the single-channel-driven power by about $\frac{3}{4}$ dB—more if the amplifier's power supply regulation is poor or the heat-sinks of the power transistors not too efficient.

MUSIC POWER

The Institute of High Fidelity (IHF) of America has evolved tests of dynamic power or music power. This is effectively interrupted sine wave testing, producing a test signal which is less demanding than the steady-state sine wave and more closely representing music waves. Experience has shown that an amplifier which will deliver, say, 10 W per channel of steady-state power will deliver about half as much again of music power, or 15 W music power per channel. However, dynamic tests are based on both channels running simultaneously with the music power in both channels summed. Thus our 10 W steady-state per channel amplifier would most likely have a music power rating of 30 W.

Since the loads used for testing amplifier power must not overheat or burn out, resistors of sufficient power rating must be used. These are not all that easy

FIG. 11.11. *Low-reading ohmmeter (early model 'Megger') being used to select a length of electric fire element for an amplifier load.*

to acquire at odd resistance values. However, one can use a length of electric fire element, adjusted to the required resistance as shown in Fig. 11.11. Here a low-reading ohmmeter is being used to measure the resistance.

POWER BANDWIDTH

With the input adjusted for the power capacity at 1 kHz, the generator frequency should be changed (amplitude remaining constant) to the low- and high-frequency ends of the spectrum to discover the frequencies where the power in the load falls by 3 dB. If this happens at, say, 20 Hz at the low-frequency end and 20 kHz at the high-frequency end, the power bandwidth would thus be given as 20 Hz to 20 kHz.

To facilitate tests of this kind the millivoltmeter should feature a decibel scale, preferably with the open portion of the scale occurring at about 75% of full-scale deflection.

SQUARE-WAVE TESTING

The setup shown in Fig. 11.8 can be used for square-wave testing, the audio generator, of course, being switched from sine wave to square wave delivery. The oscillograms in Fig. 11.12 show some classic results of this sort of testing which the author has obtained over the years. With a pure resistive load and running at 1 kHz one would not normally expect significant deviation from a true square wave form when the amplifier is in the 'high quality' or hi-fi category. A typical result is shown at (*a*). As already intimated, corner rounding is to be expected at 10 kHz, and (*b*) shows a display like that. Sometimes a little overshoot or ringing occurs even across a load of pure resistance, as shown at (*c*).

If the amplifier is very stable it should handle heavily reactive loads without undue wave mutilation, but possibly with mild overshoot as shown at (*d*). Here the load consisted of 8 ohms of pure resistance in parallel with 2 μF of capacitance. Some amplifiers react more violently with this kind of load, one example being shown at (*e*) and another at (*f*). All those with reactive loads with the exception of (*f*), which was at 1 kHz, were taken at 10 kHz. Sometimes the bandwidth of the amplifier is inadequate to accommodate the higher-order harmonics of the square wave, the output signal then tends to change towards sine wave form as shown at (*g*). This was taken at half-power at 10 kHz, the implication thus being that the power bandwidth of the amplifier is little more than up to 10 kHz. Later tests proved this to be the case.

To summarise, square wave testing provides a speedy appraisal of amplifier performance, including frequency, phase and stability. Generally speaking, an amplifier that exhibits severe overshoot or ringing tendencies sounds less pleasant than one which is less prone to such disturbances. The whole picture of square wave testing is given in Fig. 11.13. Tests are commonly made at half

(a)

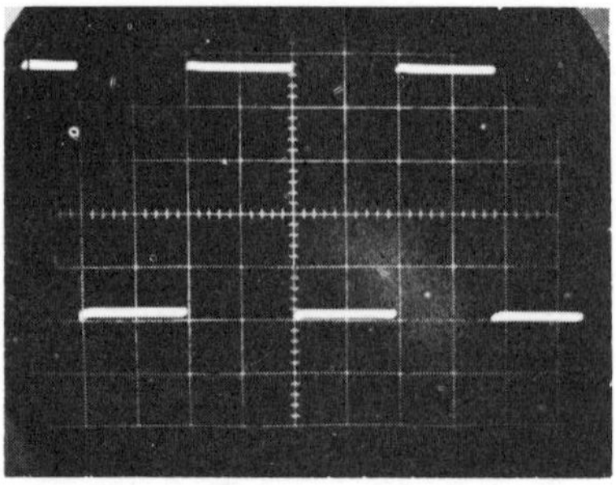

(b)

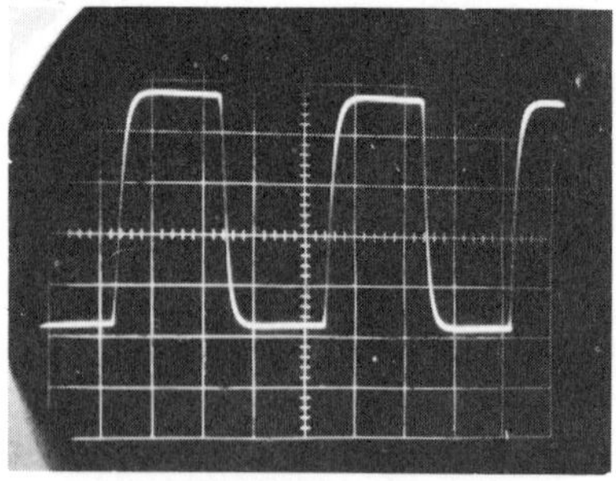

(c)

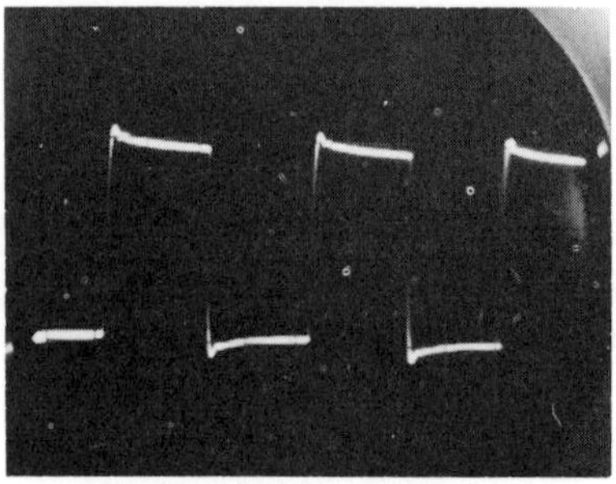

(d)

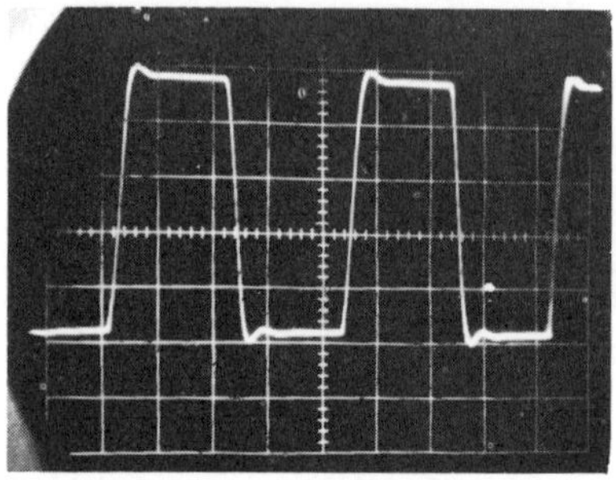

(e)

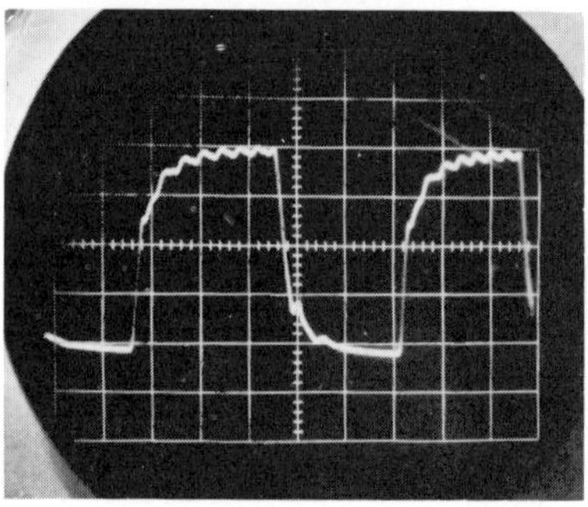

(f)

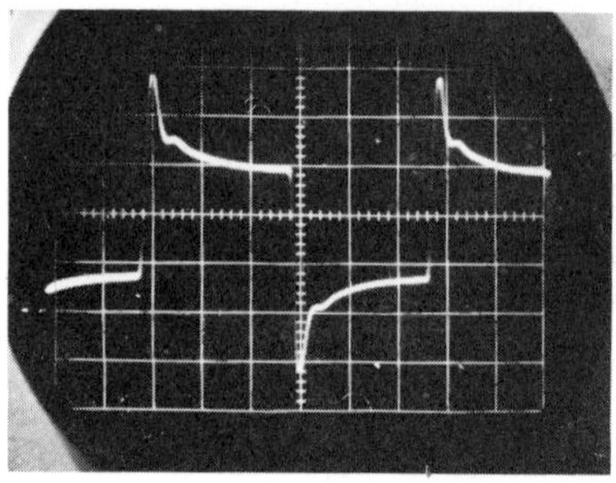

(g)

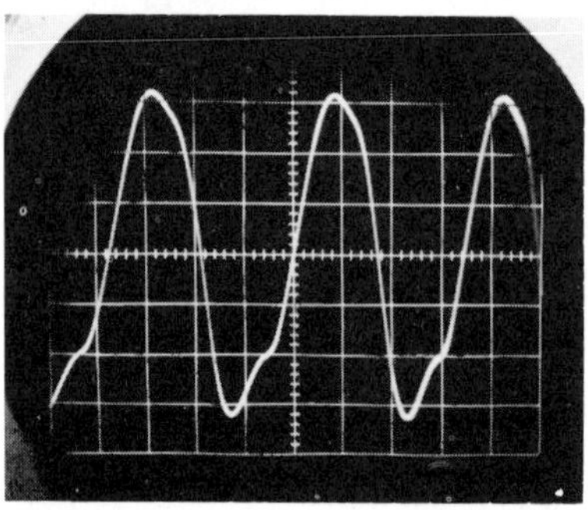

FIG. 11.12. *Oscillograms from square-wave testing. These are fully explained in the text.*

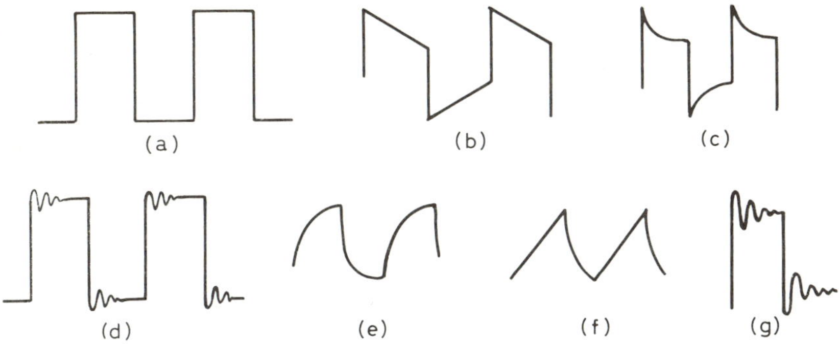

FIG. 11.13. (a) *A perfect square wave.* (b) *Its output through a normal amplifier. At* (c) *is shown poor l.f. response, at* (d) *'ringing', at* (e) *poor h.f. response, at* (f) *severely restricted response, and at* (g) *severe 'ringing', as may be caused by a mismatch in the output stage.*

sine wave steady-state power, but it is often instructive to see how power changes affect the wave shape.

NOISE TESTS

The noise performance of an amplifier is often given as 'hum and noise so many dBs below full power into a stipulated load value'. We can test in this way with the Fig. 11.8 setup. First move is to establish a dB datum on the audio millivoltmeter corresponding to maximum output power into, say, 8 ohms and at 1 kHz. The signal generator is then removed and the selected input shorted.

The noise and hum components across the load can then be related to the established dB datum by increasing the sensitivity of the millivoltmeter (e.g., by turning it to a very low range, possibly the lowest, depending on the amount of hum and noise present). Say, for example, the dB datum was established at 20 dB on the scale, and by increasing the millivoltmeter readout sensitivity by 60 dB the hum plus noise also reads 20 dB, the hum and noise is then 60 dB below the full 1 kHz power of the amplifier. Similarly, if the hum and noise reading is, say, 10 dB the ratio would be 70 dB relative to full power, and so forth.

Note that this test gives the *unweighted* dB ratio over the full spectrum of the amplifier and test instruments. When a specification gives *weighted* ratios the implication is that filters have been included in the readout circuit to provide ratios more in keeping with the 'annoyance value' of the hum and noise from the subjective point of view. Weighted ratios are larger than unweighted ones.

HARMONIC DISTORTION TESTS

The setup for measuring total harmonic distortion is given in Fig. 11.14. With S_1 bypassing the distortion test set and the amplifier adjusted as shown, the input sine wave is increased to provide the output power (see under the

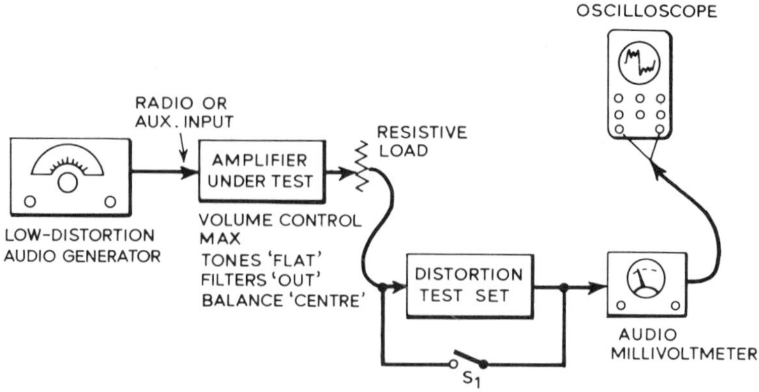

FIG. 11.14. *Setup for distortion measurement (see text).*

section dealing with power measurement on page 182) at which it is required to measure the THD.

S_1 is then opened and the distortion test set adjusted to the frequency at which it is required to measure the distortion after first having switched the instrument to 'set 100%' (see Figs. 11.5 and 11.6, for example). This merely by-passes the notch filter. The millivoltmeter should next be adjusted to the highest range consistent with the establishment of a dB datum and the power in the load, with the input attenuator of the distortion test set adjusted accordingly.

The notch filter is then brought into circuit by switching the distortion test set to 'read'. The plan is then to tune the notch by the controls on the facia to the fundamental frequency, which is indicated by a rapid fall in millivoltmeter readout. To increase the sensitivity, the millivoltmeter is switched down a range (this is commonly a 20 dB step or, the same, a voltage sensitivity increase of ten times) and the notch filter is further tuned with the coarse and fine controls until the readout drops even more. This is continued by switching the millivoltmeter down by a known number of decibels at a time until the very least readout is achieved. However, at the same time the balance controls, also on the facia, must be carefully adjusted, first the coarse and then the fine, as with the notching tuning controls, until the notch deletes all the fundamental signal and the balance controls 'balance' the remaining THD.

The readout will then be the r.m.s. voltage of the summed power of the total harmonics in the load. It will also be in part amplifier noise and hum. However, when required the hum can easily be removed by switching in the high-pass (l.f. cut) filter. For normal THD measurements at 1 kHz and above the filter is kept in circuit, but when reading distortion below 1 kHz and at low frequencies the filter might have to be switched out to secure sufficient sensitivity.

The amount of distortion is first given as a dB ratio. For example, if it is found that the distortion readout is, say, 40 dB below the readout due to the full 1 kHz power of the amplifier, then obviously the THD is 40 dB below the power at which the distortion was measured. 40 dB in voltage ratio is 100 times, and since the distortion is less than the amplifier power by that amount the ratio can be expressed as 1 %. Similarly, −60 dB THD is 0·1 %, −20 dB 10 % and so forth.

The simple THD analyser clearly works over the amplifier bandwidth and thus reads noise signals as well as distortion harmonics. Normally, the THD readout is well above the noise of the amplifier (unless, of course, the amplifier happens to possess a remarkably low distortion characteristic and not too good noise performance!). However, the noise of the amplifier can trouble the read-readout and introduce some error—making the THD appear greater than it really is—when THD is measured at low amplifier power. Normally, though, one commonly views the THD residual on an oscilloscope connected to the milli-voltmeter output (as in Fig. 11.14), so if noise is interfering with the reading this can easily be seen on the display. Noise and THD add essentially in quadrature, so by measuring the noise amplitude on the THD display it is possible to make a suitable correction—albeit not a very accurate one without a great deal of trouble.

The author has found it best to measure full power distortion with the amplifier's volume control at maximum, and then reduce the power for lower power distortion tests by turning down the volume control (keeping the generator input constant). This often improves the noise performance at the lower powers because in most amplifiers the volume control is located between the preamplifier and power amplifier sections and most noise is contributed by the early preamplifier stages. Thus by turning down the volume control the noise voltage is reduced by the same amount as the signal voltage.

When the input is passed through the preamplifier section it will be appreciated that the distortion in this section is being added (again in quadrature) to that in the power amplifier section. To minimise the distortion in the preamplifier, therefore, it is not a good thing to test at high power with the volume control turned down, since this will run the preamplifier section more towards non-linearity than necessary, with a consequent rise in THD readout.

On the other hand, the distortion in the preamplifier may have to be measured (especially to find out the distortion rise when the preamplifier is operated towards maximum output). To do this one can then turn the volume control right down and get the necessary power in the output load by turning the generator signal well up. The oscilloscope will indicate when the preamplifier section is being overloaded by waveform clipping, occurring in this case, of course, well before the full output of the power amplifier section has been reached.

PREAMPLIFIER OVERLOAD

This brings us neatly to another important test, which is, in fact, the overload performance of the preamplifier section. Important in this respect is the overload performance of the magnetic pickup preamplifier at 1 kHz. This is of particular importance because some magnetic cartridges when playing high velocity records can yield a signal voltage rising well towards 45 mV. The nominal sensitivity at the magnetic input is 3 mV. This, or a little higher (given by some authorities as 5 mV), corresponds to the 'average' output of the pickup when playing a music record. If the output is likely to rise to some 45 mV on peak velocities, therefore, the preamplifier should handle signal of that amplitude without running into severe overload or clipping.

Maximum input to such overload can be appraised by setting up as in Fig. 11.7, turning well down the amplifier's volume control and turning up the generator signal at 1 kHz until the displayed waveform shows clipping. This, of course, must happen significantly below the clipping level of the power amplifier as previously defined. Just at the clipping point the generator output signal is measured, and this corresponds to the r.m.s. value of sine wave signal up to overload. Fig. 11.15 oscillogram shows how music signal can be clipped due to a poor pickup preamplifier overload performance. Any amplifier which overloads much below about 50 mV r.m.s. at 1 kHz at the magnetic pickup

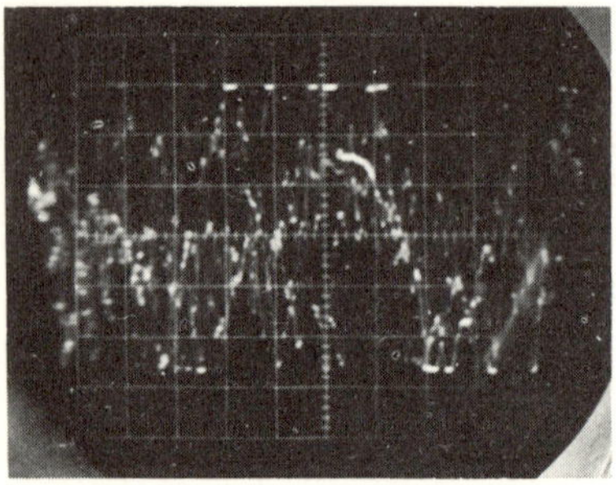

FIG. 11.15. *Clipping on music wave due to preamplifier overload.*

input has, in the author's opinion, insufficient overload margin. Some amplifiers, however, include an input preset, allowing the input sensitivity to be adjusted to suit the pickup in use in terms of dynamic range and overload.

With stereo amplifiers both channels are generally checked and when making distortion measurements it is best to take two readings, one with one channel only driven and the other with both channels driven simultaneously. This is because when both channels are driven together the supply voltage can fall and hence affect both the power output and distortion factor.

MORE ABOUT DISTORTION MEASUREMENT

In some class B type amplifiers (really towards class AB since the output stage biasing is adjusted for a smallish quiescent current) there is a tendency for the distortion to rise slightly as the power is turned down to low levels. This is a function of the design, top-flight amplifiers barely showing any increase at all, and if there is an increase it is generally of no significance since the higher power distortion is so low anyway. Nevertheless, with less exacting equipment of this kind the distortion can rise from a level that *is* significant as the power is turned down. It is commonly due to a function of the 'switching' of the output transistors and tends to increase also as the *frequency* of the test signal is increased. It is a sort of crossover distortion resulting from non-linearity of the overall transfer characteristic at the switching point. It is not present in class A amplifiers because the output transistors are not 'switched' like they are when biased for class B (AB etc.). Thus one can be pretty sure that a class A amplifier will exhibit a distortion characteristic falling fairly linearly with reducing power output.

Fig. 11.16 shows mild crossover distortion on an output sine wave. This was obtained at a power of about 100 mW at 10 kHz. Fig. 11.17 shows the nature of distortion which might appear on the oscilloscope in Fig. 11.14 after all the fundamental frequency has been deleted. The 'peaky' component is characteristic of crossover distortion, and in some instances the residual

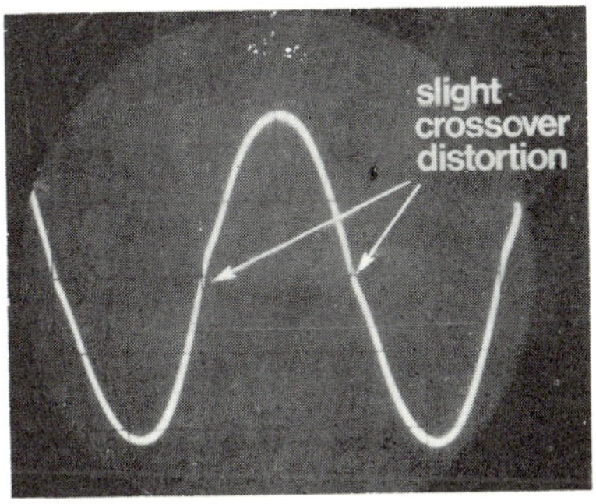

FIG. 11.16. *Mild crossover distortion at low power and 10 kHz.*

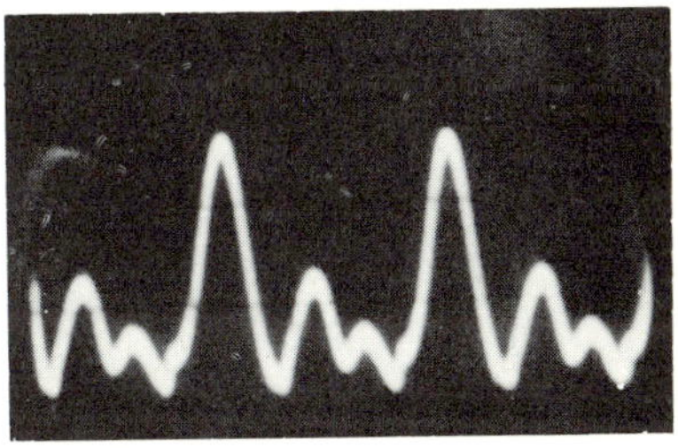

FIG. 11.17. *THD displayed on oscilloscope in Fig. 11.13 (see text).*

approaches square wave form, thus indicating the presence of high-order odd-numbered harmonics.

It will be appreciated, of course, that the distortion percentage is based on the ratio of the r.m.s. voltage of the sine wave fundamental to the r.m.s. voltage of the residual, and since the latter can deviate significantly from sine wave form the ratio is not really a true one, bearing in mind that the millivoltmeter is calibrated in terms of the r.m.s. voltage of a sine wave. Nevertheless, this is the commonly accepted way of distortion measurement. Perhaps this is one reason why there is no absolute relationship between the amount of distortion measured (assuming that the distortion is not excessive of course!) and its degree of subjective unpleasantness. There is possibly another factor here and that is since modern amplifiers have a passband extending beyond the audible frequencies crossover distortion occurring at the top of the spectrum (as in Fig. 11.16) might influence the transient nature of the music waves in a manner to produce a vague subjective unpleasantness. A wide passband is nevertheless necessary for hi-fi reproduction to maintain the correct nature of the music waves and thus to avoid impairment of the transient performance. The music waves should be maintained but not modified by the addition of spurious harmonics.

We have already noted that distortion measurement at low level is made difficult due to the signal falling towards the level of the amplifier's residual noise. This can be combated to some degree by a good wave analyser. If this has a noise equivalent bandwidth of, say, 7 Hz, then the readout relative to the noise is improved by 30 dB or so, depending on the amplifier's bandwidth. Thus while 0·1% THD can be measured down to noise with a distortion analyser looking at 100 mW output from a 10 W amplifier of 80 dB signal-to-noise ratio, a wave analyser will allow a measurement down towards 0·003% on a selected harmonic. In other words, 0·1% selected harmonic distortion from such an amplifier could be measured down to about 0·1 mW (e.g., 30 dB below 100 mW).

Another problem lies in obtaining a sufficiently low distortion test signal. It was mentioned earlier that a filter can be used to help 'purify' the signal from an audio generator. Fig. 11.18 shows how this is done. Here the tuned circuit *LC* attenuates all the harmonics, passing mostly only the frequency to which it

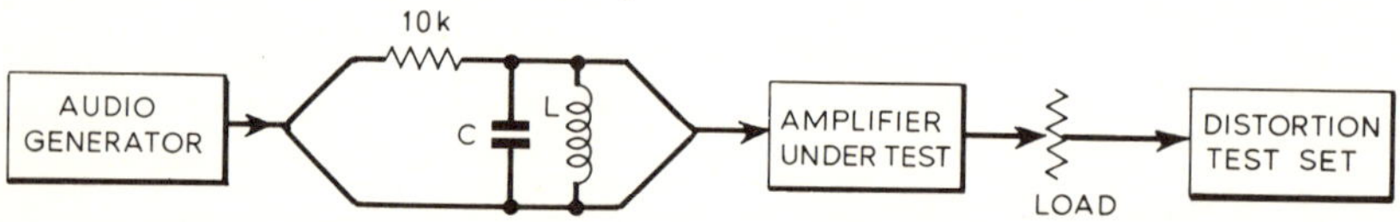

FIG. 11.18. *A filter like this between the audio generator and test amplifier reduces the test signal distortion.*

is resonated, corresponding to that tuned by the audio generator. A high-*Q* filter like this can reduce the generator's distortion to about one-fifth of its unfiltered value. To tune to 1 kHz, a typical test frequency, *L* should be 200 mH and *C* 0·13 μF. *L* should be air-cored because dust-iron cores can themselves introduce third-harmonic distortion. 200 mH can be obtained by winding 4,200 turns of 34 s.w.g. enamel-covered copper wire on to a ½ in. diameter former within a width of 1 in.

INPUT SENSITIVITY

The sensitivity at each programme input can be measured with the setup shown in Fig. 11.8. This is done at 1 kHz with the amplifier adjusted as shown. It is necessary to connect another audio millivoltmeter in parallel with the input signal and measure the input sensitivity voltage from this when the output millivoltmeter indicates the full or rated power of the amplifier.

With the J. E. Sugden Si 453 generator it is also possible to check the accuracy of the RIAA pickup equalising by switching to the 'equalised' output. This is done as for frequency response testing.

FREQUENCY RESPONSE TESTING

Again we can use the setup shown in Fig. 11.8. The output power should be adjusted to about 1 W and a dB datum at 1 kHz established on the audio millivoltmeter. It is then merely necessary to plot the dB deviation at a number of frequencies over the audio spectrum, usually from 20 Hz to 20 kHz, always ensuring, of course, that the generator signal remains at a constant amplitude at all the frequencies plotted.

OUTPUT IMPEDANCE

The output impedance of an amplifier depends on the amount and nature of the applied negative feedback, and because of capacitive couplings in the feedback loop the impedance commonly rises as the frequency reaches the lower part of the spectrum. Thus it may be at, say, 0·3 ohm at 1 kHz and above and 1 ohm at 50 Hz. The ratio of the speaker or load impedance to the output impedance of the amplifier is a measure of the amplifier's damping factor, and it is this which is usually specified. Thus a low output impedance gives a high damping factor, a factor of 16 corresponding to an output impedance of 0·5 ohm when used with a speaker of 8 ohms. A high damping factor is desirable for applying electromagnetic damping to the cone of the speaker unit.

Measurement is commonly made at 50 Hz, and the idea is to run the amplifier up to about a quarter of its rated output at 50 Hz into a specified resistive load (e.g., a load value with which the amplifier is designed to work, commonly 8 ohms with transistor amplifiers) and then measure the r.m.s. voltage across the load. Next the load is disconnected and the open-circuit output r.m.s. voltage measured. The damping factor is then calculated from $V_2/V_1 - V_2$, where V_1 and V_2 are the no-load and loaded r.m.s. voltages respectively.

Index

INDEX

INDEX